Advances in Sustainability Science and Technology

The book series aims at bringing together valuable and novel scientific contributions that address the critical issues of renewable energy, sustainable building, sustainable manufacturing, and other sustainability science and technology topics that have an impact in this diverse and fast-changing research community in academia and industry.

The areas to be covered are

- Climate change and mitigation, atmospheric carbon reduction, global warming
- Sustainability science, sustainability technologies
- Sustainable building technologies
- Intelligent buildings
- Sustainable energy generation
- Combined heat and power and district heating systems
- Control and optimization of renewable energy systems
- Smart grids and micro grids, local energy markets
- Smart cities, smart buildings, smart districts, smart countryside
- Energy and environmental assessment in buildings and cities
- Sustainable design, innovation and services
- Sustainable manufacturing processes and technology
- Sustainable manufacturing systems and enterprises
- Decision support for sustainability
- Micro/nanomachining, microelectromechanical machines (MEMS)
- Sustainable transport, smart vehicles and smart roads
- Information technology and artificial intelligence applied to sustainability
- Big data and data analytics applied to sustainability
- Sustainable food production, sustainable horticulture and agriculture
- Sustainability of air, water and other natural resources
- Sustainability policy, shaping the future, the triple bottom line, the circular economy

High quality content is an essential feature for all book proposals accepted for the series. It is expected that editors of all accepted volumes will ensure that contributions are subjected to an appropriate level of reviewing process and adhere to KES quality principles.

The series will include monographs, edited volumes, and selected proceedings.

Kapil Narula

The Climate Catalyst

Building Positive Momentum for Climate Action

Springer

Kapil Narula
Former Senior Analyst
Climate Champions Team
Dubai, United Arab Emirates

ISSN 2662-6829 ISSN 2662-6837 (electronic)
Advances in Sustainability Science and Technology
ISBN 978-981-92-5212-1 ISBN 978-981-92-5213-8 (eBook)
https://doi.org/10.1007/978-981-92-5213-8

This Springer imprint is published by the registered company Springer Nature Singapore Pte Ltd.
The registered company address is: 152 Beach Road, #21-01/04 Gateway East, Singapore 189721, Singapore

To the Supreme Soul and the divine energy that nurtures the Universe

To Mother Earth, which sustains all life

To all Souls working to protect the climate

To our shared and sustainable future

Preface

The Earth is the only planet known to humans that supports life. This is possible due to the unique conditions provided by the interaction of the atmosphere, hydrosphere, lithosphere, cryosphere, and biosphere, which together provide conditions necessary for life on Earth. However, the 'safe operating space for humanity' is increasingly being threatened by climate change and degradation of critical Earth system processes, which is amplifying environmental, economic, and social risks, with several negative consequences for sustainable development.

Climate change is emerging as an important challenge, and its impacts are being felt globally in daily life through unprecedented heat waves, rising sea levels, increased droughts, higher ice melt, changing precipitation patterns and more frequent extreme weather events. These changes are impacting the functioning of natural systems, human systems and economic systems and the risks are compounding in a non-linear manner.

Humans, as the most intelligent and evolved species on Earth, have an obligation to address climate change at a speed and scale which limits irreversible damage to the Earth system and its functioning. Climate ambition must be set at the highest possible level and climate commitments need to be translated into measurable, on-ground outcomes. This requires bridging the gap between climate ambition, commitments and real-world climate action. Scalable and replicable solutions that can be adapted across regions and sectors must be implemented globally. Climate action must also be equitable, inclusive and locally beneficial. Accelerating climate action builds resilience and brings several benefits, including economic opportunities, higher energy security, enhanced ecosystem services, environmental and social co-benefits.

This book is about solutions to address climate change and is written with the intent to catalyze climate action. Its main objective is to give the reader a comprehensive understanding of the interdisciplinary issues to act on climate change. It provides a practical understanding of implementing climate solutions across different sectors, geographies and contexts. The book balances the theoretical issues related to addressing climate change and the practical domain of climate action. It covers aspects of climate governance, climate policy, climate finance, use of markets, technology, and emerging climate solutions through the use of real-world examples to provide a functional roadmap to accelerate climate action.

The book could be used as supplementary reading material for students in climate studies as well as by researchers and academics as a handbook. The content is presented in a simple and concise manner to meet the intellectual curiosity of the inquisitive reader. While it presents a deep dive into specialized topics, the book does not claim to be a reference manual but gives the reader a clearer understanding of the ways to scale up climate solutions globally, to create a positive reinforcing 'action loop'. I hope that this book will inspire the readers to build positive momentum for accelerating climate action.

Dubai, United Arab Emirates Kapil Narula
June 2026

Competing Interests The author has no competing interests to declare that are relevant to the content of this manuscript.

Contents

About the Author

Dr. Kapil Narula is a globally recognized clean energy transition strategist and climate policy advisor with over two decades of leadership experience working with the United Nations, governments, multilateral institutions, academia and industry. He is the architect of multilateral climate frameworks, national energy strategies, and institutional mechanisms to advance global decarbonization, with expertise spanning strategic policy formulation, stakeholder orchestration, translating analytical insights into actionable roadmaps and programme delivery.

He has worked at the apex levels of international climate diplomacy on United Nations Framework Convention on Climate Change (UNFCCC) led climate processes with non-state actors, implemented decarbonization strategies and developed cross-sector policy frameworks, making him exceptionally well-suited to author this book. At the Climate Champions Team, he accelerated global climate action across power, road transport, steel, cement, buildings, hydrogen, agriculture and fertilizer sectors. He co-authored the annual 'Breakthrough Agenda Report' in partnership with the International Energy Agency, advised the Climate High-Level Champions, contributed to climate negotiations at UNFCCC Conference of the Parties (COPs), and delivered technical sessions at regional climate weeks.

His earlier roles as Adviser (Energy) to the Government of India at the NITI Aayog, where he shaped national net-zero emission goals and strategies and as Economic Affairs Officer at the United Nations Economic and Social Commission for Western Asia (UN ESCWA), where he advised Arab countries on clean energy, reflect a consistent record of translating energy and climate analysis into high-impact policy outcomes. At the University of Geneva, he led applied research and policy-relevant analysis on energy systems, climate mitigation, and sustainability, producing thought leadership that informed public-sector policy debates and international discussions. Prior to that, he held progressively senior leadership positions in the Government of India as the Head of the Electrical Engineering Department on board ships managing energy and systems operations, as the Head of Electrical Faculty in a technical training college and as a Research Fellow in a strategic think tank working on energy security and maritime sustainability.

He holds an interdisciplinary Ph.D. degree in Development Economics with a focus on energy, environment and climate change, a master's degree in engineering and has a certification in sustainability and climate risk. He is an accredited Technical Expert Reviewer for Article 6 and Biennial Transparency Report by the UNFCCC. He is the author/editor of seven books, including the 'Maritime Dimension of Sustainable Energy Security (Springer, 2019) and his academic papers have appeared in *Energy, Energy Policy, Renewable and Sustainable Energy Reviews, Applied Energy* (Elsevier), *Strategic Analysis* (Routledge) and in other international forums.

Abbreviations

°C	Degree Centigrade
AAU	Assigned Amount Unit
ACCU	Australian Carbon Credit Unit
ACE	Action for Climate Empowerment
ACR	American Carbon Registry
ADB	Asian Development Bank
ADIA	Abu Dhabi Investment Authority
AEF	Agreed Electronic Format
AFD	Agence Française de Développement
AfDB	African Development Bank
AFOLU	Agriculture, Forestry and Other Land Use
AHP	Analytical Hierarchy Process
AI	Artificial Intelligence
AIB	Association of Issuing Bodies
ALS	Airborne Laser Scanning
API	Application Programming Interface
APR	Association of Plastic Recyclers
AQI	Air Quality Index
AR	Assessment Report
ARC	Attributes of Recycled Content
ASI	Aluminium Stewardship Initiative
ATJ	Alcohol-To-Jet
AWS	Amazon Web Services
BAT	Best Available Technology
BCCTF	Bangladesh Climate Change Trust Fund
BCM	Bromochloromethane
BECCS	Bioenergy With Carbon Capture and Storage
BESS	Battery Energy Storage System
BEV	Battery Electric Vehicle
BF-CCUS	Blast Furnace Carbon Capture, Utilization and Storage
Bio-SR	Smelting Reduction with Biomass

BMS	Building Management System
BMZ	Federal Ministry for Economic Cooperation and Development (English translation)
BNG	Biodiversity Net Gain
BPA	Bonneville Power Administration
BSA	Biodiversity Stewardship Agreement
BTR	Biennial Transparency Report
BU	Biodiversity Unit
BVRio	Bolsa verde do Rio de Janeiro
BYF	Belite-Ye'elimite-Ferrite
C3S	Copernicus Climate Change Service
CAAF	Conference on Aviation Alternative Fuels
CAD	Climate Action Data
CAES	Compressed Air Energy Storage
CAFÉ	Corporate Average Fuel Economy
CAGR	Compound Average Growth Rate
CAISO	California Independent System Operator
CAP	Credit Aqua Positive
CapEx	Capital Expenditure
CAPMF	Climate Actions and Policies Measurement Framework
CAR	Climate Action Reserve
CARB	California Air Resources Board
CARP	Centralized Accounting and Reporting Platform
CatSim	Catastrophe Simulation
CAV	Connected and Autonomous Vehicle
CBA	Cost-Benefit Analysis
CBAM	Carbon Border Adjustment Mechanism
CBDR-RC	Common But Differentiated Responsibilities and Respective Capabilities
CCA	Climate Change Authority
CCAC	Climate and Clean Air Coalition
CCAI	Climate Change AI
CCAP	Climate Change Action Plan
CCB	Climate, Community and Biodiversity
CCE	Coalition for Climate Entrepreneurship
CCM	Circular Credits Mechanism
CCMM	CIF Capital Markets Mechanism
CCRIF	Caribbean Catastrophe Risk Insurance Facility
CCS	Carbon Capture and Storage
CCT	Cirrus Cloud Thinning
CDM	Clean Development Mechanism
CDP	Carbon Disclosure Project
CDR	Carbon Dioxide Removal
CDS	Climate Data Store
CEE	Certificats D'économies D'énergie

CEEU	CORSIA Eligible Emissions Unit
CEF	CORSIA Eligible Fuels
CEFS	Community Energy Financing Scheme
CER	Certified Emission Reduction
CFC	Chlorofluorocarbon
CFE	24/7 Carbon-Free Energy
CFLI	Climate Finance Leadership Initiative
CGE	Computable General Equilibrium
CH_4	Methane
CHLC	Climate High-Level Champion
CIE	Climate Impact Explorer
CIF	Climate Investment Funds
CII	Carbon Intensity Indicator
CIP	Climate Investment Platform
CLCPA	Climate Leadership and Community Protection Act
CLIMADA	Climate Adaptation
CMA	Conference of the Parties Serving as the Meeting of the Parties to the Paris Agreement
CMP	Conference of the Parties Serving as the Meeting of the Parties to the Kyoto Protocol
CO_2	Carbon Dioxide
CoC	Chain of Custody
CoO	Certificates of Origin
COP	Conference of the Parties
CORDEX	Coordinated Regional Climate Downscaling Experiment
CORSIA	Carbon Offsetting and Reduction Scheme for International Aviation
CPCB	Central Pollution Control Board
CSA	Climate-Smart Agriculture
CSP	Concentrated Solar Power
CSR	Corporate Social Responsibility
CSRD	Corporate Sustainability Reporting Directive
CTCN	Climate Technology Centre and Network
CTF	Clean Technology Fund
Cumac	Cumulated and Discounted
CVC	Corporate Venture Capital
D3	Third Derivative
DAC	Direct Air Capture
DBSA	Development Bank of Southern Africa
DCS	Data Collection System
DESNZ	Department for Energy Security and Net Zero
DFC	Development Finance Corporation
DiD	Difference-In-Differences
DMFC	Direct Methanol Fuel Cell
DNA	Designated National Authority

DOC	Direct Ocean Capture
DOE	Department of Energy
DPG	Digital Public Good
DPI	Digital Public Infrastructure
DPSP	Dedicated Private Sector Program
DRCM	Dynamic Reference Cover Modeling
DRR	Disaster Risk Reduction
E/MSY	Extinctions Per Million Species Per Year
EAC	Energy Attribute Certificate
EAD	Environment Agency—Abu Dhabi
EAE	Energy Access Explorer
EBRD	European Bank for Reconstruction and Development
ECCC	Environment and Climate Change Canada
EEA	European Environment Agency
EEDI	Energy Efficiency Design Index
EEXI	Energy Efficiency Existing Ship Index
EIA	Environmental Impact Assessment
EMDE	Emerging Markets and Developing Economy
eMRV	Ecological MRV
En-ROADS	Energy Rapid Overview and Decision Support
EPA	Environmental Protection Agency
EPC	Energy Performance Certificate
EPR	Extended Producer Responsibility
EPS	Environmental Policy Stringency
ERGaR	European Renewable Gas Registry
ERS	Electric Road Systems
ERU	Emission Reduction Unit
ERW	Enhanced Rock Weathering
ESCO	Energy Service Company
ESD	Education for Sustainable Development
ESDS	Earth Science Data System
ESG	Environmental, Social and Governance
ESRS	European Sustainability Reporting Standard
ETF	Enhanced Transparency Framework
ETS	Emission Trading Scheme
EU	European Union
EV	Electric Vehicle
FAIR	Fundamental AI Research
FAR	First Assessment Report
FCDO	Foreign, Commonwealth and Development Office
FCV	Fuel Cell Vehicle
FFEM	French Facility for Global Environment (English translation)
FIT	Feed-In Tariff
FMCG	Fast-Moving Consumer Goods
FMCP	Facilitative, Multilateral Consideration of Progress

FOAK	First-Of-A-Kind
FONERWA	Rwanda Green Fund (English translation)
FRLD	Fund for Responding to Loss and Damage
FSC	Forest Stewardship Council
FSD	Full Self-Driving
FT	Fischer-Tropsch
FTC	Financial, Technology Development and Transfer, and Capacity-Building
GC	Granular Certificate
GCAA	Global Climate Action Agenda
GCCA+	Global Climate Change Alliance+
GCF	Green Climate Fund
GDP	Gross Domestic Product
GEC	Green Electricity Certificate
GEDI	Global Ecosystem Dynamics Investigation
GEE	Google Earth Engine
GEF	Global Environment Facility
GFANZ	Glasgow Financial Alliance For Net Zero
GFCS	Global Framework for Climate Services
GFI	GHG Fuel Intensity
GFOI	Global Forest Observation Initiative
GHG	Greenhouse Gas
GHSL	Global Human Settlement Layer
GIF	Generation IV International Forum
GIS	Geographic Information System
GLM	Grazing Land Management
GMP	Global Methane Pledge
GNN	Graph Neural Network
GO	Guarantees of Origin
GPFG	Government Pension Fund Global
GPS	Global Positioning System
GPU	Graphics Processing Unit
GRI	Global Reporting Initiative
GST	Global Stocktake
Gt	Gigaton
GWP	Global Warming Potential
H2-DRI- EAF	Hydrogen-Direct Reduced Iron- Electric Arc Furnace
H2-DRI	Hydrogen Direct Reduced Iron
H2-GO	Hydrogen Guarantees of Origin
HALEU	High-Assay Low-Enriched Uranium
HBFC	Hydrobromofluorocarbon
HCFC	Hydrochlorofluorocarbon
HEFA	Hydroprocessed Esters and Fatty Acid
HFC	Hydrofluorocarbon
HTGR	High Temperature, Gas-Cooled Reactor

HTR-PM	High Temperature Gas-Cooled Reactor—Pebble-Bed Module
HVDC	High Voltage Direct Current
HWR	Heavy Water Reactor
IB	Issuing Body
IBRD	International Bank for Reconstruction and Development
ICAO	International Civil Aviation Organization
ICCTF	Indonesia Climate Change Trust Fund
ICF	International Climate Finance
ICP	Internal Carbon Pricing
IDA	International Development Association
IDB	Inter-American Development Bank
IEA	International Energy Agency
IET	International Emissions Trading
IFAD	International Fund for Agricultural Development
IFC	International Finance Corporation
IF-CAP	Innovative Finance Facility for Climate in Asia and the Pacific
IIX	Impact Investment Exchange
IKI	International Climate Initiative
IMD	India Meteorological Department
IMO	International Maritime Organization
INDC	Intended Nationally Determined Contribution
IoT	Internet of Things
IPCC	Intergovernmental Panel on Climate Change
IPG	International Partners Group
IPO	Initial Public Offering
I-REC(E)	International Renewable Energy Certificate for Electricity
I-REC	International Renewable Energy Certificate
IRENA	International Renewable Energy Agency
ISCC	International Sustainability and Carbon Certification
IsDB	Islamic Development Bank
ISSB	International Sustainability Standards Board
ITMO	Internationally Transferred Mitigation Outcome
IUCN	International Union for the Conservation of Nature
JBIC	Japan Bank for International Cooperation
JETP	Just Energy Transition Partnership
JI	Joint Implementation
JICA	Japan International Cooperation Agency
KPI	Key Performance Indicator
K-REC	Korea Renewable Energy Certificate
kW	Kilowatt
kWh	Kilowatt-Hour
LC3	Limestone Calcined Clay Cement
LCAF	Lower Carbon Aviation Fuel
LCFS	Low Carbon Fuel Standard
LCIPP	Local Communities and Indigenous Peoples Platform

LCOE	Levelized Cost of Electricity
LDC	Least Developed Country
LDCF	Least Developed Countries Fund
LGC	Large-Scale Generation Certificate
LiDAR	Light Detection and Ranging
LINKS	Local and Indigenous Knowledge System
LLDC	Landlocked Developing Country
LNG	Liquefied Natural Gas
LPAA	Lima-Paris Action Agenda
LTA	Long Term Agreement
LT-LEDS	Long-Term Low Greenhouse Gas Emission Development Strategy
LULUCF	Land Use, Land-Use Change, and Forestry
LWR	Light Water Reactor
M&A	Mergers and Acquisitions
MACC	Marginal Abatement Cost Curve
MB	Methyl Bromide
MCA	Multi Criteria Analysis
MCB	Marine Cloud Brightening
mCDR	Marine Carbon Dioxide Removal
MDB	Multilateral Development Bank
MEPS	Marine Environment Protection Committee
MIGA	Multilateral Investment Guarantee Agency
ML	Machine Learning
MMA	Ministry of Environment and Climate Change
MMQR	Method of Moments Quantile Regression
MOB	Methane-Oxidizing Bacteria
MOCCAE	Ministry of Climate Change and Environment
MODIS	Moderate Resolution Imaging Spectroradiometer
MOE	Molten Oxide Electrolysis
MoEFCC	Ministry of Environment, Forest and Climate Change
MOF	Metal-Organic Framework
MOX	Mixed Oxide
MPGCA	Marrakech Partnership for Global Climate Action
M-RETS	Midwest Renewable Energy Tracking System
MRV	Monitoring, Reporting, and Verification
MWh	Megawatt-Hour
N_2O	Nitrous Oxide
NAFCC	National Adaptation Fund for Climate Change
NAP	National Adaptation Plan
NAPA	National Adaptation Programme of Action
NAZCA	Non-State Actor Zone for Climate Action
NbS	Nature-Based Solution
NCF	National Climate Fund
NCQG	New Collective Quantified Goal
NDC	Nationally Determined Contribution

NDF	Nordic Development Fund
NEAC	National Environment Awareness Campaign
NEDO	New Energy and Industrial Technology Development Organization
NF_3	Nitrogen Trifluoride
NFC	Non-Fossil Certificate
NICFI	Norway International Climate and Forest Initiative
NLP	Natural Language Processing
NORAD	Norwegian Agency for Development Cooperation
NO_X	Nitrogen Oxides
NVZ	Nitrate Vulnerable Zone
O&M	Operations and Maintenance
OAE	Ocean Alkalinity Enhancement
OBI	Orange Bond Initiative
OBP	Ocean-Bound Plastic
ODA	Official Development Assistance
ODS	Ozone-Depleting Substances
OECD	Organisation for Economic Co-Operation and Development
OTEC	Ocean Thermal Energy Conversion
PACM	Paris Agreement Crediting Mechanism
PCM	Phase Change Material
PCX	Plastic Credit Exchange
PE	Private Equity
PEM	Proton Exchange Membrane
PES	Payments For Ecosystem Services
PFC	Perfluorocarbon
PHEV	Plug-In Hybrid Electric Vehicle
PIBO	Producers, Importers, and Brand Owners
PJM GATS	Pennsylvania, New Jersey, and Maryland Generation Attribute Tracking System
PLA	Polylactic Acid
PoS	Proof of Sustainability
PPF	Project Preparation Facility
ppm	Parts per million
PPP	Public-Private Partnership
PPRS	Plastic Pollution Reduction Standard
PRISM	Platform for Real-Time Impact and Situation Monitoring
PSC	Perovskite Solar Cell
PSF	People's Survival Fund
PSH	Pumped Storage Hydropower
PtX	Power-to-X
PV	Photovoltaic
PWR	Pressurized Water Reactor
PWRS	Plastic Waste Reduction Standard
QA/QC	Quality Assurance/Quality Control

QELRC	Quantified Emission Limitation and Reduction Commitment
QGIS	Quantum Geographic Information System
R&D	Research and Development
RBCF	Results Based Climate Finance
RCP	Representative Concentration Pathway
RE	Renewable Energy
REC	Renewable Energy Certificate
RED	Renewable Energy Directive
REDD	Reducing Emissions from Deforestation and Forest Degradation
REGO	Renewable Energy Guarantees of Origin
RFNBO	Renewable Fuels of Non-Biological Origin
RFS	Renewable Fuel Standard
RGGI	Regional Greenhouse Gas Initiative
RGGO	Renewable Gas Guarantees of Origin
RMI	Rocky Mountain Institute
RMS	Recycled Material Standard
RNG	Renewable Natural Gas
RPS	Renewable Portfolio Standard
RSB	Roundtable on Sustainable Biomaterials
RTK	Revenue Tonne Kilometer
RUSLE	Revised Universal Soil Loss Equation
SAF	Sustainable Aviation Fuel
SAFc	Sustainable Aviation Fuel Certificate
SAI	Stratospheric Aerosol Injection
SAPCC	State Action Plans on Climate Change
SATAN	Stratospheric Aerosol Transport and Nucleation
SBI	Subsidiary Body for Implementation
SBSP	Space Based Solar Power
SBSTA	Subsidiary Body for Scientific and Technological Advice
SBTi	Science Based Targets Initiative
SCC	Soil Carbon Credit
SCCF	Special Climate Change Fund
SCF	Strategic Climate Fund
SCM	Supplementary Cementitious Material
SCoPEx	Stratospheric Controlled Perturbation Experiment
SCS	Sustainable Constructional Steel
SDC	Swiss Agency for Development and Cooperation
SDG	Sustainable Development Goal
SECO	State Secretariat for Economic Affairs
SEEMP	Ship Energy Efficiency Management Plan
SEforALL	Sustainable Energy for All
SEPAL	System for Earth Observations, Data Access, Processing and Analysis for Land Monitoring
SF_6	Sulphur Hexafluoride
SFDR	Sustainable Finance Disclosure Regulation

SIA	Social Impact Assessment
SIDS	Small Island Developing State
SLCP	Short-Lived Climate Pollutant
SLM	Sustainable Land Management
SME	Small and Medium-Sized Enterprise
SMES	Superconducting Magnetic Energy Storage
SMR	Small Modular Reactor
SOC	Soil Organic Carbon
SOE	State Owned Entity
SOFC	Solid Oxide Fuel Cell
SOP	Standard Operating Procedure
SPV	Special Purpose Vehicle
SREC	Solar Renewable Energy Certificate
SRM	Solar Radiation Modification
SSP	Shared Socioeconomic Pathway
STC	Small-Scale Technology Certificate
SWF	Sovereign Wealth Fund
TCFD	Task Force on Climate-Related Financial Disclosures
T-EAC	Time-Based Energy Attribute Certificate
TER	Technical Expert Review
TGR-BF	Top Gas Recycling Blast Furnace
TLS	Terrestrial Laser Scanning
toe	Tonne of Oil Equivalent
TPA	Tonnes Per Annum
TRL	Technology Readiness Level
TWh	Terawatt-Hour
UCO	Used Cooking Oil
UDB	Union Database
UNCED	United Nations Conference on Environment and Development
UNDP	United Nations Development Programme
UNEP	United Nations Environment Programme
UNFCCC	United Nations Framework Convention on Climate Change
UNGC	United Nations Global Compact
USAID	U.S. Agency for International Development
UUV	Unmanned Underwater Vehicle
V2G	Vehicle-To-Grid
VAT	Value Added Tax
VBC	Voluntary Biodiversity Credit
VC	Venture Capitalist
VCM	Voluntary Carbon Market
VCS	Verified Carbon Standard
VGF	Viability Gap Funding
VOC	Volatile Organic Compound
WBG	World Bank Group
WCC	World Climate Conference

WCP	World Climate Programme
WLB	Women's Livelihood Bonds
WQT	Water Quality Trading
WREGIS	Western Renewable Energy Generation Information System
WUE	Water Use Effectiveness
WWF	World Wildlife Fund
YoY	Year-on-Year

List of Figures

List of Tables

Chapter 1
Climate Change: Challenges, Impacts and Actions

Abstract This introductory chapter introduces the global challenge of climate change. It characterizes the complexity of the climate challenge, provides scientific evidence, and builds the case for climate action. Trends of global emissions by sector, gases and countries are presented, along with the socio-economic drivers of emissions in the backdrop of the remaining carbon budget. The chapter presents the impact of climate change on natural systems, human systems and economic systems to highlight the all-encompassing nature of the climate challenge. At the end, it briefly discusses the overarching actions for addressing climate change under the broad heads of mitigation, adaptation and resilience. The chapter concludes that the time for transformative action on climate change is not in the distant future, but now.

Keywords Climate change · Greenhouse gas · Climate impacts · Mitigation · Adaptation · Resilience

1.1 Introduction

Climate change refers to long-term shifts in temperatures and weather patterns which can be due to natural or man-made causes. It is a major concern for countries, regions, cities, businesses and individuals as it affects ecosystems that support life and biodiversity. A changing climate has several consequences with impacts beyond the environment. This includes economic, social and cultural effects on humans and physical impacts on flora and fauna. The drivers of climate change and its impacts are rapidly growing. As the Earth is a globally connected system, there are additional risks due to the multiplicative and cascading effects on the physical system. Addressing climate change is therefore extremely important and actions must be undertaken by the international community, governments, and people at scale and speed to maintain a livable climate.

This chapter introduces the global challenge of climate change and characterizes the complexity of the challenge. It provides the scientific evidence of climate change and builds the case for action. Section 1.3 presents the trends of global emissions

K. Narula, *The Climate Catalyst*, Advances in Sustainability Science and Technology,
https://doi.org/10.1007/978-981-92-5213-8_1

by sector, gases and countries and establishes the connection between emissions and global temperature rise. It also presents socio-economic drivers of emissions and climate forecasts in the backdrop of the remaining carbon budget. Section 1.4 presents the impact of climate change on natural systems, human systems and economic systems to highlight the all-encompassing nature of the challenge. Section 1.5 briefly discusses the overarching actions for addressing climate change under the broad heads of mitigation, adaptation and resilience before concluding.

1.2 Climate Change as a Global Challenge

1.2.1 An Existential Threat

Climate change has been characterized as a 'direct existential threat' by the UN Secretary General [7]. It poses a risk of systemic failure of human civilization as it has a widespread impact on human wellbeing and planetary health. Increasing risks include unprecedented warming of atmosphere and oceans; extreme weather events such as severe heatwaves, frequent floods, droughts, and wildfires; ocean acidification threating marine ecosystems; increased melting of polar ice and sea level rise resulting in water inundation in coastal areas with high risk of submergence of low-lying small island nations; increasing impacts on human health witnessed in rising diseases and heat-related deaths; increasing global competition and conflicts over natural resources triggered by food and water stress; ecosystem and biodiversity loss causing mass mortalities and undermining vital ecosystem services as evidenced by the risk of extinction to over 1 million species [8]. The disruptions to natural ecosystems, degradation of planetary health and collapsing biodiversity could lead to cascading impacts across environmental, social and economic domains. This may eventually disrupt the existing functioning of human society and its capacity for recovery in the future.

These impacts have been highlighted in the *Sixth Assessment Report of the Intergovernmental Panel on Climate Change* [16] and the *Lancet Countdown on health and climate change* [29]. Climate change and resulting environmental risks continue to be highest ranked in terms of severity of impact in *The Global Risks Report 2025* [37].

1.2.2 A 'Wicked' Problem

Wicked problems were characterized to describe social policy planning challenges where traditional scientific problem-solving methods were proving to be inadequate [28]. Wicked problems are fundamentally different due to the characteristic of the problem itself. Societal problems such as climate change have no definitive formulation, as the understanding of the cause of the problem changes with shifting timeframes and impacts. They have complex causes which propagate through multiple interconnected factors across economic, political, social, and physical systems with feedback loops that are difficult to isolate. There is no clear endpoint where the problem can be considered as definitively 'solved,' and continuous action is required to manage the problem.

In a pluralistic society, there is no undisputable public good and in a wicked problem there is disagreement on what constitutes a viable solution, as different stakeholders have conflicting perspectives based on their values and self-interests. The framing of the problem also influences the range of options of proposed solutions. There are several viable solutions to the problem, but there is no exhaustive list of all possible solutions. There is no ultimate test of a solution as effects may be long-term or difficult to measure, and remedial actions may have potentially irreversible consequences. Policies that respond to wicked problems cannot be declared as right or wrong, and there may be no definitive and objective answers.

Climate change has several of the above characteristics and can be considered as a 'wicked problem'. It has a dynamic and complex nature that has ecological, economic, social, and environmental dimensions, crosses political boundaries, involves multiple stakeholders and countries with competing interests, impacts the entire society and has several solutions which work in specific contexts.

1.2.3 The Scientific Evidence

The *State of the Global Climate Report* by the World Meteorological Organization [39] presents an updated status of seven key climate indicators. The following trends have been identified in recent years.

(a) Greenhouse gas concentration in the atmosphere: In 2023, the atmospheric concentration of carbon dioxide, methane and nitrous oxide reached the highest levels in the last 800,000 years. The concentration of carbon dioxide (CO_2) in the atmosphere was 420.0 ± 0.1 parts per million (ppm), which corresponds to 3276 Gt CO_2 in the atmosphere. The latest update reports the concentration of carbon dioxide (CO_2) to be about 424 ppm in 2024 [40].

(b) Global surface temperature: The year 2024 was the warmest year in the 175-year observational record, and each of the past ten years, 2015–2024, was individually the ten warmest years on record. The annually averaged global mean near-surface temperature in 2024 was 1.55 ± 0.13 °C above the 1850–1900 average (used to represent pre-industrial conditions). 2025 is on track to be the second or third warmest year on record.

(c) Ocean heat content: Oceans absorb about 90% of excess heat in Earth's climate system, and their warming is an important indicator of long-term climate shifts. In 2024, ocean heat content reached the highest level in the 65-year observational record, and each year over the past eight years has set a new record for ocean heat content. The rate of ocean warming over the past two decades, 2005–2024, is more than twice that observed over 1960–2005. This warming is irreversible for centuries to millennia, and is projected to continue through the twenty-first century and beyond.

(d) Sea-Level rise: Global sea levels continue to rise due to thermal expansion and melting ice caps, which is a clear indicator of climate change impacts. In 2024, global mean sea level reached a record high and increased by about 110 mm from 1993. The long-term rate of sea-level rise has more than doubled, increasing from 2.1 mm per year (1993–2002) to 4.7 mm per year (2015–2024). A small drop has been observed in 2025 compared to 2024, likely due to natural climate variability.

(e) Ocean acidification: Absorption of CO_2 in oceans leads to a decline in ocean pH, which has decreased from 8.11 to 8.045 over the past four decades. While there are regional variations, globally, ocean surface pH has changed at a rate of -0.017 ± 0.001 pH units per decade over the period 1985–2023. Deep-ocean pH changes are irreversible for centuries to millennia, and ocean acidification is projected to continue to increase in this century.

(f) Glacier mass balance: Glacier mass loss from 2021–2022 to 2023–2024 represents the most negative three-year glacier mass balance on record. Seven of the ten most negative annual glacier mass balances since 1950 have occurred since 2016, demonstrating a continuing trend of accelerated glacier mass loss in the 2020s. Glaciers lost a record 450 Gt of ice from October 2023 to September 2024, which is equivalent to 1.2 mm of global mean sea-level rise and is more than double the average annual loss rates of the late twentieth century.

(g) Arctic and Antarctic sea ice extent: The extents of sea-ice in the Antarctic and Arctic regions in 2024 were both below their respective 1991–2020 averages throughout the annual cycle. The minimum daily extent of Arctic sea-ice in 2024 was the seventh lowest (1979 to present), and all 18 minima have occurred in the past 18 years. The minimum daily extent of sea-ice in the Antarctic region in 2024 was the second lowest minimum and marked the third consecutive year that minimum Antarctic sea-ice extent dropped below 2 million km. Arctic sea-ice extent in March 2025 was the lowest annual maximum and Antarctic sea-ice extent has remained well below average throughout 2025 [40].

1.2.4 Tipping Elements and Planetary Boundaries

A tipping point is a critical threshold that, when crossed, leads to large, accelerating and often irreversible changes in the system. In the context of climate change, tipping elements are large-scale components of the Earth system that can lead to abrupt, non-linear, and potentially irreversible changes to the climate on human timescales, once thresholds are reached. These are also characterized by self-amplifying feedback loops which can destabilize the climate system and may have cascading effects, triggering other tipping elements leading to systemic risks.

Initially, nine 'core' elements which contributed substantially to Earth system functioning, and seven 'regional' tipping elements were identified [1]. The nine 'core' global tipping elements with the likely threshold and importance are shown in Table 1.1.

The regional elements include Barents Sea ice, Sahel and West African monsoon, boreal forest—Southern die-off and northern expansion, labrador/Irminger seas

Table 1.1 Global tipping elements

Tipping element and likely threshold	Description	Importance
Greenland ice sheet, 1.5 °C (0.8–3 °C)	Massive ice sheet up to three-kilometer-thick covering Greenland -all year round	Irreversible melting could raise sea levels by up to 7 m, affecting coastal cities worldwide
West Antarctic ice sheet, 1.5 °C (1–3 °C)	Marine-based ice sheet in West Antarctic	Collapse could add over 3 m to global sea levels
East Antarctic subglacial basins, 3 °C (2–6 °C)	Deep ice basins (e.g., Totten glacier) resting below sea level	Potential long-term sea-level rise if destabilized
Atlantic meridional overturning circulation, 4 °C (1.4–8 °C)	Large-Scale Ocean current system including the Gulf stream responsible for the mild climate of northwestern Europe	Disruption could shift global temperature and precipitation patterns, harming agricultural output and cooling in the North Atlantic
Arctic winter sea ice, 6.3 °C (4.5–8.7 °C)	Floating sea ice that forms each winter with a thickness of a few meters	Loss reduces reflectivity, accelerating Arctic and global warming
Mountain glaciers	Glaciers in mountain ranges such as the Andes and Himalayas which provide freshwater	Loss threatens water security and contributes to sea-level rise
Boreal permafrost thawing, 1.5 °C (1–2.3 °C)	Permanently frozen ground across northern latitudes estimated to store about a thousand billion tons of carbon in the upper three meters of the frozen soil	Thawing releases methane and CO_2, triggering a feedback loop that accelerates warming could potentially lead to large-scale boreal permafrost collapse
Amazon rainforest, 3.5 °C (2–6 °C)	Largest tropical rainforest which plays a critical role in the water and carbon cycles	Dieback could lead to releasing carbon and altering rainfall patterns
Low-latitude coral reefs, 1.5 °C (1–2 °C)	Coral ecosystems in tropical and subtropical waters influencing marine food chains, nutrient and carbon cycles in the ocean	Highly sensitive to temperature. Bleaching could affect marine biodiversity and fish stocks increasing food insecurity

(subpolar gyre), East Antarctic ice sheet (specific basins) and mountain glaciers. Several other tipping elements have been included over time based on evolving scientific evidence and consensus [22].

Tipping elements are closely linked with the 'Planetary Boundaries' framework, which identifies nine Earth system processes that maintain planetary stability, defining humanity's 'safe operating space'. Crossing these boundaries can dramatically increase the risks for triggering tipping elements. The 2025 assessment concluded that seven of the nine boundaries are transgressed, as shown in Fig. 1.1, increasing the risks to trigger global processes that impact the stability and resilience of the Earth. Planetary boundaries are interconnected through complex Earth system processes. Therefore, these systems cannot be addressed in isolation and all nine boundaries must be respected to preserve humanity's safe operating space.

The nine planetary boundaries and their status are enumerated in Table 1.2.

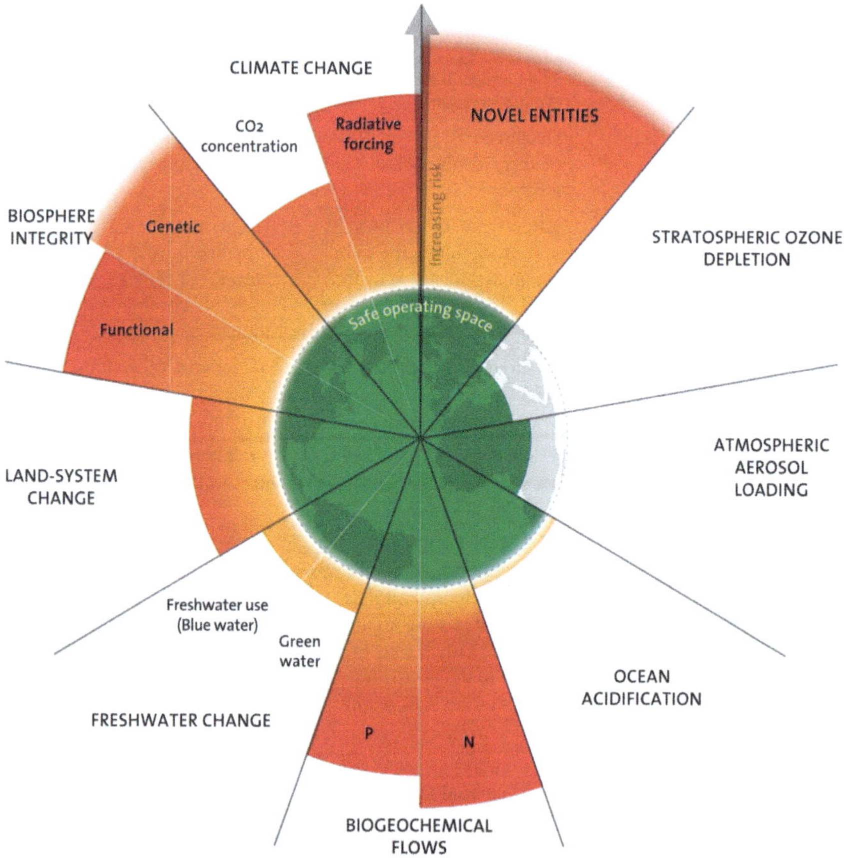

Fig. 1.1 Planetary boundary status 2025. *Source* Stockholm Resilience Centre [31]

Table 1.2 Description and status of planetary boundaries. *Data source* Stockholm Resilience Centre [31]

Planetary boundary	Description and status
Climate change	The climate change boundary measured by atmospheric CO_2 concentration and radiative forcing is transgressed their safe thresholds
Biosphere integrity	The loss of genetic diversity and the decline in the functional integrity of the biosphere are outside safe levels
Land-system change	Tracks the conversion of forests and natural lands to cropland. Globally, the remaining forest areas in tropical, boreal, and temperate biomes have fallen below safe levels which diminishes ecological functions like carbon sequestration and moisture recycling
Freshwater change	Monitors consumptive freshwater use. Human-induced disturbances of rivers and lakes and soil moisture have exceeded the planetary boundary
Biogeochemical flows	Tracks disruption of nitrogen and phosphorus cycles. As excessive fertilizer use has disrupted global biogeochemical flows
Ocean acidification	Assessed via carbonate saturation in seawater, which impacts marine life.
Atmospheric aerosol loading	Measures particulate matter from human activities and natural sources, which influences the climate by altering temperature and precipitation patterns. Current global averages remain within the safe operating space, but this balance masks severe regional impacts
Stratospheric ozone depletion	Protects life by blocking harmful UV radiation. A slow recovery is happening and it is in the 'safe operating space'
Novel entities	Introduction of synthetic chemicals, plastics, heavy metals, which may intervene in evolutionary processes and change the functioning of the earth system. This boundary is already exceeded

1.2.5 The Urgency for Action

As is evident from the above sections, the global climate system is undergoing rapid transformation. The *Intergovernmental Panel on Climate Change (IPCC) Assessment Report (AR) 6* findings have confirmed that human activities have unequivocally resulted in global warming primarily through the emission of Greenhouse Gases (GHG), rapidly changing the Earth's climate [21]. The report has consistently emphasized that the window for effective climate action is rapidly closing and the consequences of delaying climate action may lead to crossing of scientifically established warming thresholds, like 1.5 or 2 °C above pre-industrial levels.

This may lead to runaway warming, which directly translates to severe extreme weather events. Many climate-related risks have been revised upward, and the projected long-term impacts have been assessed to be multiple times higher than those observed currently. Climate risks, their adverse impacts and related losses and damages are projected to escalate, and scientific evidence indicates that every additional 0.5 °C of global temperature rise will lead to 'clearly discernible increases in the frequency and severity of heat extremes, heavy rainfall events and regional droughts'. It is further forecasted that climatic and non-climatic risks will increasingly interact, creating compound and cascading risks that are more complex and difficult to manage.

IPCC AR6 highlights that some future changes are unavoidable and/or irreversible due to the existing temperature rise, but the likelihood of abrupt and/or irreversible changes increases with higher temperatures. As seen from the continuing increase of atmospheric CO_2 concentration in the last decade and with 2024 being the first year where the global mean temperature was 1.55 °C above the pre-industrial level (1850–1900 average), this is not a distant future scenario but a present reality [41]. The accelerating trend in the rate of change of climate indicators increases the threat. The impact of climate change is non-linear, which implies that the probability of low-likelihood outcomes associated with very large adverse impacts increases with higher global warming levels.

Earth's climate is a complex system which is deeply interconnected with many feedback loops. Degradation of one component of the climate system amplifies problems in other components, leading to a systemic change. As an example, melting glaciers contribute to sea-level rise, which then exacerbates coastal flooding and saltwater intrusion in low-lying areas. Ocean warming and acidification together degrade healthy marine ecosystems, impacting marine biodiversity and food security. This interconnectedness exacerbates vulnerabilities, leading to compounding adverse effects on natural ecosystems. These impacts of climate change amplify threats to human societies and economies and the interconnectedness leads to a spiraling nature of these impacts, triggering further vulnerabilities.

Overwhelming scientific evidence collectively paints an unequivocal picture of urgency as global warming is accelerating at an unprecedented pace, leading to widespread and intensifying impacts, including a growing likelihood of triggering non-linear threats and irreversible planetary changes. This underscores the need for immediate and integrated, systemic responses to avoid continuous, escalating impacts.

It is now well established through scientific observations and forecasts that the impact of changing climate on the planet has already been locked in. Further, climate change is irreversible on human timescales, which has a significant implication for the future generations. The long-term impact implies that even if global emissions are immediately stopped, future impacts are already locked in for generations due to the persistence of GHGs in the atmosphere.

Hence, decisions made today will profoundly shape the planet for countless future generations, which creates a moral imperative for immediate action to address climate change. An increase of every additional fraction of a degree of warming exacerbates irreversible changes and underscores the urgency to break these reinforcing loops before they lead to long-lasting planetary alterations. Large-scale action to reduce GHG emissions must be immediate, and the longer we wait to reduce emissions, the more challenging it will become to limit global warming to safe levels.

1.3 Global Emissions

The greenhouse effect occurs naturally due to absorption of infrared radiation emitted by the Earth's surface by water vapour, carbon dioxide, clouds, other gases and aerosols. The Earth's atmosphere traps the reflected heat, which warms the planet, making it habitable. However, increasing concentration of GHGs from anthropogenic emissions contributes to an additional radiative forcing which enhances the magnitude of this effect. The changes to the natural greenhouse effect, called the enhanced greenhouse effect, are the cause of additional warming.

1.3.1 Greenhouse Gases

Greenhouse gases (GHGs) are gaseous constituents of the atmosphere that absorb and emit radiation, causing the greenhouse effect. Water vapor is the most significant GHG, which accounts for about half of the greenhouse effect [23]. Human activity has almost no direct impact on its concentration, and it is very short-lived in the atmosphere (lifespan of hours to days), unlike most GHGs. However, water vapor is more effective in trapping heat in the atmosphere, and its increased quantities (higher precipitation due to climate change) can amplify the impact of other GHGs, due to the presence of feedback mechanisms.

Global Warming Potential (GWP) is a standard metric used to compare the impact of different GHGs over a specific time horizon such as 20, 100, or 500 years [14]. It provides a base unit (CO_2-equivalent) to compare, assess and aggregate the warming effects of various gases, despite their different impacts on radiative forcing. The main GHGs, their GWP, atmospheric lifetime and anthropogenic sources are shown in Table 1.3.

Chlorofluorocarbons, Chlorocarbons, Hydrochlorocarbons, Bromocarbons, Hydrobromocarbons Halons, Halogenated Alcohols, Ethers, Furans, Aldehydes and Ketones are some of the additional GHGs and substances that have different GWPs and are found in relatively lower amounts.

1.3.2 Global Emissions Trends

Global GHG emissions have increased significantly since 1900, primarily due to the use of fossil fuels. GHG emissions grew rapidly to 59 ± 6.6 GtCO2-eq in 2019 [18]. Emissions continue to rise, but the rate of growth has slowed down.

Table 1.3 GHGs and their GWP. *Data source* Greenhouse Gas Protocol [6]

Greenhouse gas (GHG)	GWP (100-year time horizon, as per AR6)	Atmospheric lifetime	Main anthropogenic sources
Carbon dioxide (CO_2)	1 (Reference)	100–1000 Years	Fossil fuel combustion, deforestation, industry, land-use change
Methane (CH_4) (differs for non-fossil and fossil)	27–30	~12 Years	Agriculture (livestock, rice production), fossil fuel extraction, landfills
Nitrous oxide (N_2O)	273	~121 Years	Agriculture (fertilizer use), livestock manure, fossil fuel combustion, industrial processes
Nitrogen trifluoride (NF_3)	17,400	~740 Years	Electronics manufacturing (e.g., LCD panels, solar cells, semiconductors)
Sulphur hexafluoride (SF_6)	24,300	~3200 Years	Electrical insulation (switchgear), magnesium and aluminum production
Hydrofluorocarbons (HFCs)	100–12,000+	~15–40 Years (varies by type)	Refrigeration, air conditioning, foam blowing agents, aerosol propellants.
Perfluorocarbons (PFCs)	6630–11,100	Thousands of years	Aluminum production, electronics manufacturing (e.g., CF_4, C_2F_6)

1.3.2.1 GHG Emissions by Sector

As shown in Fig. 1.2, global GHG emissions grew from approximately 34 Gt $CO_{2\ eqvt.}$ in 1990 to about 51 Gt $CO_{2\ eqvt.}$ in 2023. In 2023, the energy sector (39 Gt) was the largest contributor to GHG emissions, followed by agriculture (6 Gt), industrial processes (3 Gt), waste (2 Gt) and Land Use, Land-Use Change, and Forestry (LULUCF).

Within the energy sector, electricity and heat production from power plants are the main sources of emissions. Emissions from different modes of transportation (cars, trucks, aviation, shipping) and fuel combustion in industry for manufacturing, mining, construction and factories also account for a large share of energy emissions. GHG emissions also result from energy use for heating, cooling, cooking, lighting in residential and commercial buildings. Other emission sources in the energy sector include use of fuel for non-energy uses such as chemical feedstock and refining. Emissions under the agriculture sector occur from crop cultivation, methane from

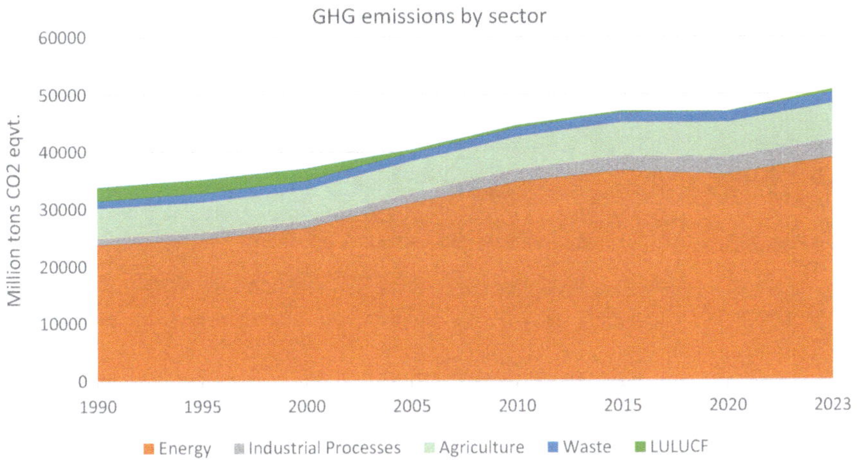

Fig. 1.2 GHG emissions by sector. *Data source* Climate Watch [3]. *Notes* Data is different from the estimates by IPCC and some other global databases

livestock (enteric fermentation and manure), rice cultivation and use of fertilizers. Forestry and land use changes include emissions from deforestation, peatland degradation and forest fires and together with agriculture are often referred to as emissions from the Agriculture, Forestry and Other Land Use (AFOLU) sector. Emissions from industrial processes include CO_2 emissions in the production of clinker, production of ammonia and during other chemical processes. Emissions from the waste sector include solid waste disposal (landfills), wastewater treatment, incineration of organic matter, which produces methane and nitrous oxide.

1.3.2.2 GHG Emissions by Gases

Figure 1.3 shows the GHG emissions from different gases. F-gases include hydrofluorocarbons, nitrogen trifluoride, perfluorocarbons, sulphur hexafluoride and other fluorinated compounds. Hydrofluorocarbons represent around 90% of all F-gases.

CO_2 was the largest GHG, which contributed to about 37 Gt, followed by methane (10 Gt), Nitrous oxide (2.7 Gt) and F gases (1.4 Gt). Methane emissions are linked to oil and natural gas extraction, coal mining and natural gas flaring and leaks, agriculture and waste sector. The majority of nitrous oxide emissions are from the agriculture sector, resulting from high fertilizer use. GHG emissions from F-gases, which have a high GWP (used in specialized industrial applications), are also growing rapidly.

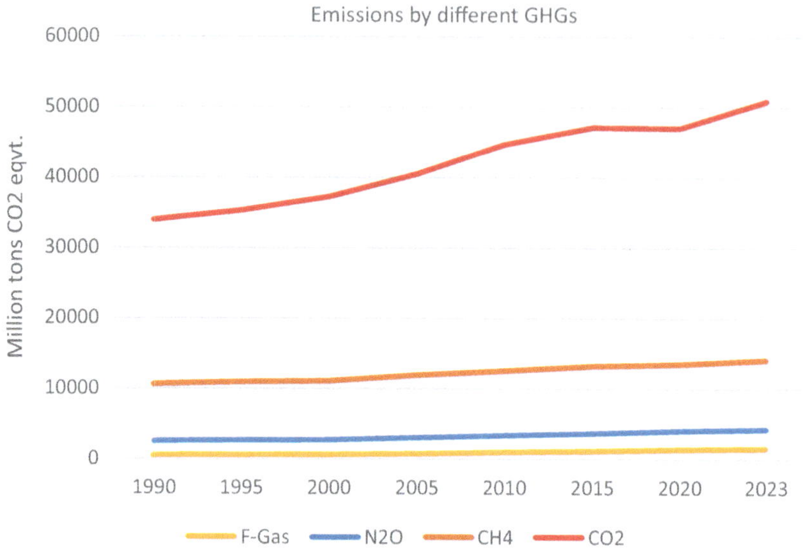

Fig. 1.3 Emissions by major GHGs. *Data source* Climate Watch [3]. *Notes* Data includes emissions from LULUCF

1.3.2.3 GHG Emissions by Countries

Figure 1.4 shows the ten largest GHG emitter countries in the past few decades. The top 10 countries contributed to about 60% of global GHG emissions while the 100 least-emitting countries contributed less than 3%. The top contributors to global GHG emissions in 2023 were China (13 Gt), the United States (5.4 Gt), India (4 Gt), Brazil (2 Gt), Indonesia (1.5Gt), Russia, Iran, Japan, Canada and Saudi Arabia. All other countries together contributed to about 19 Gt, which was about the same as the top two contributors. China overtook the United States to become the world's largest GHG emitter in the mid-2000s, and in 2023 it contributed to about one-fourth of the global emissions.

1.3.2.4 Cumulative GHG Emissions

It is estimated that about 2.7 trillion tons of CO_{2e} have been emitted into the atmosphere since 1850 [4]. Cumulative emissions are important to understand because GHGs persist in the atmosphere for centuries and global warming is the result of the accumulated stock of GHG emissions since industrialization.

Figure 1.5 shows the total cumulative emissions of CO_2 from 1750–2024 by the top 10 emitting countries. Only emissions from fossil fuels and industry are accounted for. The global CO_2 emissions were about 1.8 trillion tons, with the US contributing to almost a quarter of the emissions. Until the mid-twentieth century, more than half

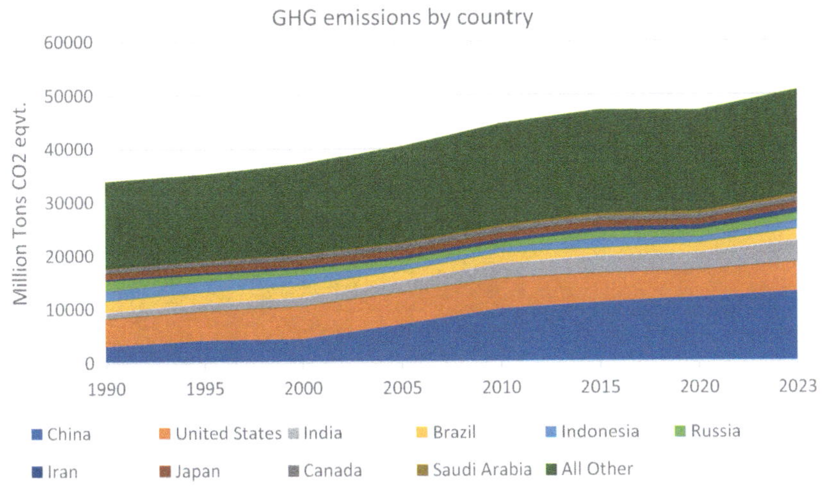

Fig. 1.4 Ten largest GHG emitter countries. *Data source* Climate Watch [3]. *Notes* Data is based on emissions produced within a country's borders

of these emissions were emitted by Europe, with the major contributor being the UK, while China and India contributed to less than 1% each.

1.3.2.5 GHG Emissions and Temperature Rise

There is a direct causal relationship between GHG emissions and temperature rise. Accumulated GHG emissions result in increased concentration of gases in the atmosphere, which can be measured in parts per million (PPM). The concentration of CO_2 in pre-industrial times is considered the baseline and is estimated as 280 ppm. By the end of 2024, the estimated global average concentration of CO_2 was about 424 ppm and it continues to be on a rising trend [38]. Each 1 ppm increase in CO_2 represents approximately 2.13 Gt of carbon or 7.82 Gt of CO_2 in the atmosphere.

There is a near-linear correlation between cumulative GHG emissions in the atmosphere and higher temperatures. It is estimated that every 100 ppm increase in CO_2 concentration leads to approximately a 1 °C rise in global mean temperature. The 144 ppm increase between 2024 and pre-industrial time closely corresponds to the observed increase in the global average surface temperature of 1.44 °C (± 0.13 °C) above the 1850–1900 average, which is considered a reference for the pre-industrial period [42].

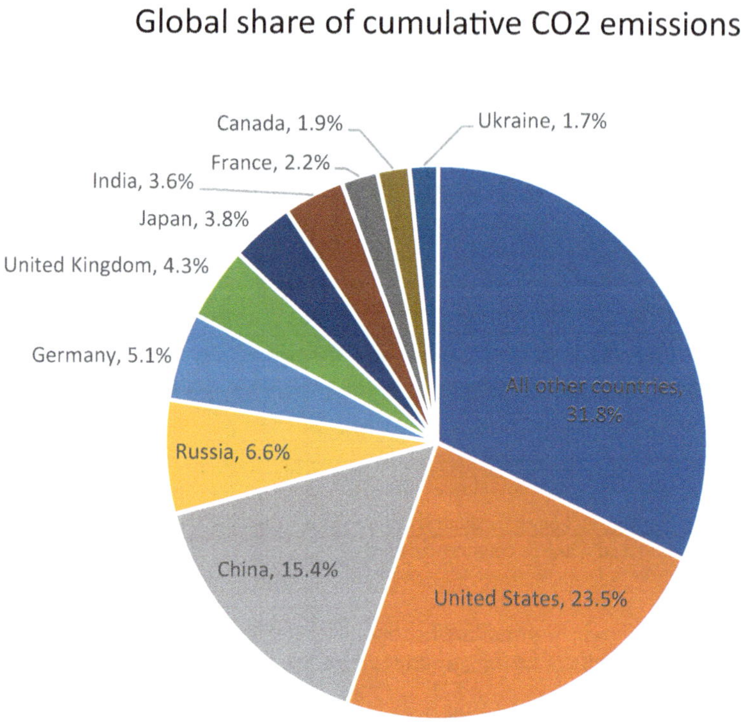

Fig. 1.5 Global share of cumulative CO_2 emissions (1750–2024) from top 10 countries. *Data source* Climate Watch [3]. *Notes* Data shows CO_2 emissions and does not cover other GHGs. Emissions from international shipping and aviation are not included as they are not attributable to countries

1.3.3 Socioeconomic Drivers of Emissions

1.3.3.1 Population Growth

Population growth is the main factor driving global GHG emissions. A larger population results in higher demand for food, energy, transportation, housing and consumable goods, which results in higher GHG emissions.

The growth in global population and the decadal growth rates from 1900 to 2020 are shown in Table 1.4. The total global population in 2025 is expected to be about 8.23 billion [35]. It is further projected that global population will continue to grow for the next 5–6 decades, peaking at approximately 10.3 billion by the mid-2080s, following which it is expected to gradually decline to around 10.2 billion by 2100 [33]. This increase in population will continue to drive energy use and emissions, unless the world rapidly switches to cleaner sources of energy.

Table 1.4 Decadal population growth. *Data source* Worldometers [43]

Year	Population (billions)	Decadal growth rate (%)
1900	1.60	–
1910	1.75	9.38
1920	1.86	6.29
1930	2.07	11.29
1940	2.30	11.11
1950	2.49	8.26
1960	3.02	21.29
1970	3.70	22.52
1980	4.45	20.27
1990	5.33	19.78
2000	6.17	15.77
2010	7.02	13.77
2020	7.89	12.39

1.3.3.2 Urbanization

Urbanization also leads to higher energy consumption as lifestyle changes are associated with increased use of motorized transport, higher consumption patterns and frequent use of electrical appliances, leading to higher per capita energy consumption. Urban infrastructure development such as commercial and residential buildings, roads, and utilities such as electricity and water also leads to increased GHG emissions as there continues to be heavy reliance on fossil fuels for providing these services.

There are large inequalities between urban and rural areas. The average per capita carbon footprint in urban areas in developing countries tends to be two to nine times larger than that of rural inhabitants [36]. Stylized estimates of per capita GHG emissions ($tCO_{2eqvt.}$/year) in rural and urban areas in countries at different stages of development are shown in Table 1.5.

Today, 55% of the world's population lives in urban areas, and this share is expected to increase to 68% by 2050 [32]. This will lead to an increase in energy use and higher per capita GHG emissions as urban areas have a comparatively higher GHG emission footprint.

Table 1.5 Estimated per capita GHG emissions at different stages of development

Region/type	Per capita GHG emissions ($tCO_{2\ eqvt.}$/year)
High-income city (car-dependent)	15–20
Rural area in developed country	3–6
Large city in developing country	1–3
Rural area in developing country	0.5–2

1.3.3.3 Economic Patterns

Economic development of a country is a major factor driving energy consumption and GHG emissions. As seen in Fig. 1.6, the per capita GHG emissions vary significantly amongst regions.

North America region is the had the highest per capita emissions in 1990 with over 21 tons $CO_{2\ eqvt}$. Which decreased to about 17 tons $CO_{2\ eqvt}$. in 2023. In 2023, East Asia and Pacific, Europe and Central Asia, Latin America and Caribbean, and Middle East and North Africa regions had higher per capita emissions than the global average of 6.3 tons $CO_{2\ eqvt}$. Small Island Developing States, South Asia and Sub-Saharan Africa had much lower per capita emissions, but these are growing to catch up with the world average. Within the regions, countries have significantly different emissions, with a trend of falling per capita emissions in developed countries such as the Organisation for Economic Co-operation and Development (OECD) and an increasing trend in emerging economies.

Higher Gross Domestic Product (GDP) is correlated with higher industrial production, as energy is provided from high-emitting fossil fuels such as coal. This is clearly observed in the case of GDP and emissions growth in China between 1973 and 2023. China's per capita GDP increased nearly 75 times while its per capita GHG emissions grew roughly by over 5 times from 1.5 to over 10 tCO_2e/year. However, on a per capita basis, it is still lower than many wealthy countries such as Qatar and heavy emitters like the US.

Economic growth in turn requires higher energy use for increasing industrial production, processing, manufacturing, transportation and other services. For example, steel production results in the emission of approximately 2 tons of CO_2 per ton of steel produced and is a major component of infrastructure development. The demand for transportation is also high in wealthier countries where people have

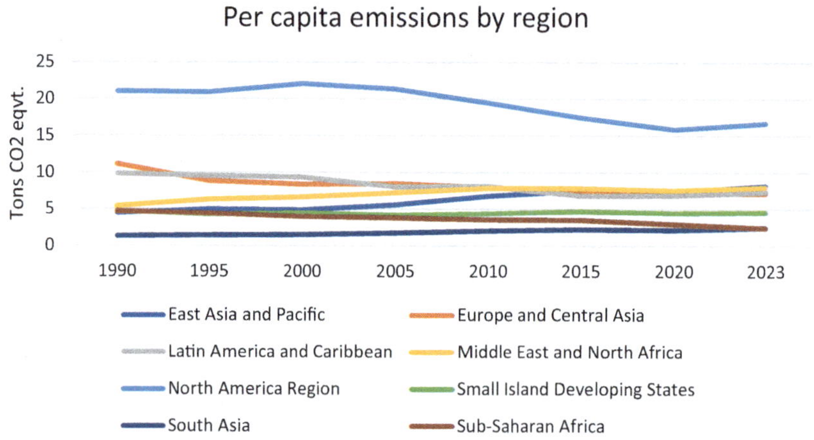

Fig. 1.6 Per capita GHG emissions in different regions. *Data source* Climate Watch [3]. *Notes* Data includes LULUCF

higher car ownership, easier and relatively cheaper availability of flights and greater demand for goods. This leads to high per capita emissions from use of personal transportation, as seen in OECD countries. Higher incomes are generally accompanied by a resource-intensive lifestyle. Some examples include larger homes equipped with heating and cooling, larger food waste and higher discretionary consumer purchases. This leads to higher per capita GHG emissions in developed countries as compared to developing countries.

1.3.4 Climate Forecasts

1.3.4.1 Climate Modeling

Climate models use the physical science knowledge to undertake earth system modeling. Computer simulations are used to model the behavior of the atmosphere, oceans, land and cryosphere to create climate forecasts which include temperature, precipitation and other physical impacts. Integrated Assessment Models (IAMs) simulate the interactions between physical systems and human systems, which include energy, economic and policy choices to estimate how climate may evolve in the future. These models are heavily dependent on assumptions about the socio-economic drivers such as population growth, technology, and policy.

Shared Socioeconomic Pathways (SSPs) are narratives of projected socioeconomic global changes up to 2100 [10]. These alternative scenarios use estimated data for population, urbanization and economic growth, based on which GHG emission scenarios are derived under different climate policies. Representative Concentration Pathways (RCPs) are also used to model climate change. These are trajectories which describe future radiative forcing (amount of radiation trapped in the atmosphere measured in watts per square meter) and GHG concentrations (measured in parts per million) which may occur in the future [11].

The Sixth Assessment Report by IPCC used a combination of different scenarios of SSPs and RCPs (radiative forcing of 1.9–8.5 W/m^2) to model future climate. The combination reflects a wide range of GHG emissions which correspond to a temperature rise of 1–5.7 °C by 2100.

1.3.4.2 Emissions Forecasts

IPCC forecasts that global surface temperatures could rise by up to 4.4 °C by 2100 in a very high GHG emissions scenario (relative to 1850–1900) or could be around 2.7 °C for an intermediate GHG emissions scenario [20]. Recent assessments indicate that current global policies may lead to about a 2.6 °C temperature increase above pre-industrial levels. More stringent targets and actions could limit this increase to about 2.2 °C above pre-industrial levels by the end of the century [2].

1.3.4.3 Remaining Carbon Budget

A total of 2770 $GtCO_2$ has been emitted since 1850, which has led to a temperature rise of about 1.2–1.3 °C since the start of the Industrial Revolution [13]. It is estimated that the remaining carbon budget to limit global warming to three different temperature targets of 1.5 °C, 1.7 °C and 2 °C (with 50% likelihood) is 170 $GtCO_2$, 525 $GtCO_2$ and 1055 $GtCO_2$, respectively [4]. This carbon budget will be exhausted within 4, 12 and 25 years from 2026 at the current rate of annual emissions. This is shown in Fig. 1.7.

1.4 Impacts of Climate Change

The escalating impacts of climate change, such as higher extreme weather events, increasing sea levels and ecosystem degradation, are already being felt globally. Climate risks are increasing at a faster rate than previously estimated, and are a threat to food, water and livelihood security. The key impacts of climate change and the accompanying risks are summarized in Table 1.6.

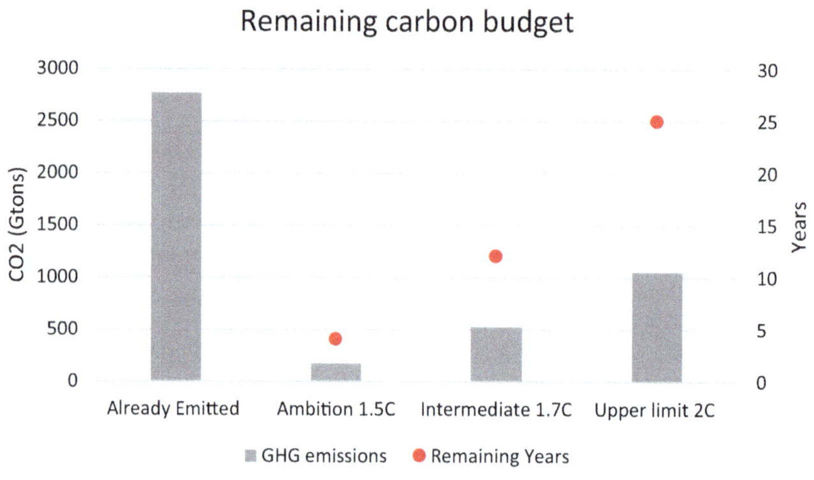

Fig. 1.7 Cumulative emissions and remaining carbon budget

Table 1.6 Climate change impacts and risks *Data source* IPCC [15]

Impacts/risks	Highlights
Observed impacts	• Widespread adverse impacts on natural and human systems • Impacts will be felt across sectors and geographies • Most vulnerable people are disproportionately affected
Vulnerability and exposure	• 3.3–3.6 Billion people live in highly vulnerable contexts • High vulnerability of several species • Vulnerability is driven by socioeconomic development and inequity
Near-term risks (2021–2040)	• More frequent and intense extreme weather events • Near-term actions cannot eliminate all projected losses and damages • High probability of multiple risks to ecosystems and humans
Mid to long-term risks (2041–2100)	• For 127 key risks, assessed mid- and long-term impacts are multiple times higher than those observed currently • Impacts escalate with an increase in every fraction of temperature • Risks depend on near-term mitigation and adaptation actions
Complex and compound risks	• Risks are becoming increasingly complex • Climatic and non-climatic risks interact and compound • Risks cascade across sectors and regions
Overshoot impacts	• Temporarily exceeding 1.5 °C creates additional severe risks • Some impacts will be irreversible even if warming is reduced • Severity depends on magnitude and duration of overshoot

1.4.1 Impacts on Natural Systems

1.4.1.1 Terrestrial Ecosystems and Biodiversity

Climate change is altering terrestrial ecosystems with significant impacts on biodiversity. According to the *Global Assessment Report on Biodiversity and Ecosystem Services*, around 1 million plant and animal species are currently threatened with extinction [9]. This includes 40% of amphibians, about 33% of corals, more than 33% of marine mammals, and 10% of all insects. Much of this can be attributed to human impact on the environment. The threat to biodiversity is also flagged by the International Union for the Conservation of Nature (IUCN), which classifies nearly 4000 species as critically endangered—species that face an extremely high risk of extinction in the wild. Current extinction rates of about 100 E/bMSY (extinctions per million species per year) are likely a thousand times higher than the background rate of 0.1 E/MSY [27].

Increasing temperatures lead to habitat fragmentation in terrestrial ecosystems and wild species are forced to migrate towards higher altitudes, often exceeding their ability to adapt. Increases in the frequency of extreme events like droughts and wildfires destroy habitats, leading to increased extinction risk, especially for vulnerable species. Changes in precipitation also affect vegetation, which may lead to desertification or altered greening with reduced biodiversity. This significantly reshapes the planet's terrestrial ecosystems and impacts biodiversity [16].

1.4.1.2 Freshwater Systems and Water Resources

Freshwater systems and potable water resources are highly vulnerable to climate change, and can significantly impact natural ecosystems. Only about 2.5% of Earth's water is classified as freshwater, but a large part of it is not accessible for drinking as it is locked away in ice caps, glaciers and underground reservoirs. Changes in rainfall patterns, increased frequency and duration of droughts and disruption in water flow patterns impact lake levels and recharge of groundwater. Warmer water temperatures lower oxygen levels in lakes, which impacts aquatic species and allows the growth of harmful algal blooms. Changes in the timing of snowmelt impact seasonal water availability, which affects freshwater systems. The impact of climate change has been observed in many regions, which has damaged the ecological integrity of rivers, lakes, and wetlands.

1.4.1.3 Ocean Ecosystems and Marine Biodiversity

Ocean ecosystems are undergoing rapid transformations due to an increase in average ocean temperature and acidification from carbon dioxide absorption, which lowers the pH of seawater. This is leading to large-scale coral bleaching, which decreases coral growth and structural integrity and is threatening marine organisms. Deoxygenation from nutrient runoff, especially from agriculture, leads to a decrease in oxygen levels in the ocean. This may lead to the creation of 'dead zones' where dissolved oxygen levels in the water are significantly reduced. Many marine organisms cannot survive in low-oxygen conditions, leading to habitat loss and declines in marine populations that can unbalance marine food webs with impacts on local marine biodiversity. Some notable examples of dead zones include the Gulf of Mexico, parts of the Baltic Sea and areas in the Gulf of Oman. Warmer ocean temperatures also increase ocean stratification, which leads to less mixing of water, thereby reducing oxygen replenishment in deeper waters. This impacts marine life and biodiversity, disrupting marine ecosystems. Rising sea levels also impact coastal ecosystems, such as mangroves and salt marshes, lowering the quality of ecosystem services that they provide. Together, these are altering the distribution of marine species through lowering reproduction rates and increasing mortality across specific marine habitats.

1.4.1.4 Critical Biomes at Risk

Some important biomes that are ecologically extremely important are facing extreme risks as they are highly sensitive to changes. Coral reef ecosystems are highly threatened by climate change as a warming ocean increases thermal stress, which contributes to coral bleaching and infectious disease [25]. Sea level rise may lead to sedimentation runoff, which can lead to the smothering of coral. Stronger and more frequent storms can cause the destruction of reef structure, while changes in precipitation can lead to increased runoff of freshwater and land-based pollutants, which contribute to algal blooms. Altered ocean currents may lead to changes in connectivity and temperature regimes that contribute to a lack of food for corals and hamper dispersal of coral larvae. *The Status of Coral Reefs of the World: 2020* reports that between 2009 and 2018 there was a progressive loss of about 14% of the coral from the world's coral reefs, primarily caused by recurring, large-scale coral bleaching events [5]. A UNEP report predicts that by 2034, severe bleaching is expected to occur annually, which may lead to widespread collapse unless corals can adapt to higher temperatures [34].

The Arctic is a critical ecosystem due to its unique role in climate regulation and its biodiversity. However, it is warming over twice as fast as the global average, leading to rapid sea ice loss and severe ecosystem changes. Several areas in the Arctic regions showed increased ocean phytoplankton blooms. Rising sea levels have led to inundation of terrestrial permafrost surrounding the Arctic Ocean, resulting in nearly 2.5 million km^2 of subsea permafrost that is at risk of thawing. The 2023 circumpolar average peak tundra greenness was the third highest in the 24-year satellite record. Changes in salmon abundance and size have also been observed, which are associated with climatic changes in freshwater and marine ecosystems [24].

Rainforests such as those in the Amazon, Congo and Southeast Asian countries have unparalleled biodiversity, housing over half of the world's terrestrial species. They act as major carbon sinks, absorbing large amounts of CO_2 and regulate atmospheric moisture, influencing regional and global rainfall patterns [16]. Rising temperatures and changing precipitation patterns can lead to stress on trees, impact tree mortality and reduce their ability to absorb CO_2. Shifts in rainfall patterns can disrupt the balance of the ecosystem, increase the risk of wildfires and can trigger shifts to savanna-like ecosystems.

The IPCC warns that losing these critical biomes could trigger far-reaching ecological and climate consequences worldwide.

1.4.2 Impacts on Human Systems

1.4.2.1 Food Security and Agricultural Productivity

Changes in rainfall patterns, higher temperatures, and extreme weather events pose an increasing threat to agricultural productivity through reduced crop yields. Changing climatic conditions can lead to the spread of pests and diseases, while higher temperatures can shrink the duration of the growing season in some regions, impacting crop production. This can lead to volatility in agricultural commodity prices and impact food security, which increases risks of famine. Higher CO_2 concentrations can reduce the protein content of crops, which lowers their nutritional value.

Climate change affects livestock through increases the risk of disease outbreaks. High temperatures can cause heat stress, leading to reduced feed intake resulting in growth deficiencies and decreased milk and meat production. Indirect impacts from changes in rainfall patterns and temperature can impact the quantity and quality of forage and feed crops, or reduced nutritional value.

1.4.2.2 Water Security and Hydrological Cycles

Climate change disrupts natural hydrological cycles, affecting water availability. Increased evaporation leads to changes in precipitation intensity, which lead to more frequent and severe floods, which may contaminate water sources. On the other hand, droughts reduce surface and groundwater availability, resulting in crop failures. This can lead to lower water security in some regions, impacting availability of water for agriculture and drinking. Retreating glaciers lead to reduced fresh water flows, while rising sea levels can lead to saltwater intrusion into coastal aquifers, rendering freshwater unusable. These impacts exacerbate existing water stress, particularly in arid and semi-arid regions, and can pose a significant threat to water security, which can lead to displacement and conflicts.

1.4.2.3 Human Health

Climate change impacts human health through both direct and indirect pathways. Direct impacts include higher heat-related mortality and morbidity during more frequent and intense heatwaves and can exacerbate respiratory illness like asthma. Indirect pathways include increased risk of infectious diseases due to an increase in the geographical range and season for vector-borne diseases (e.g., malaria, dengue fever). Water and foodborne diseases can increase due to contaminated water supplies after floods, while air quality can worsen considerably with increased wildfires. Mental health issues such as psychological distress, anxiety, and depression may also increase existing health inequities [29].

1.4.2.4 Infrastructure and the Built Environment

Climate change poses escalating risks to infrastructure and the built environment. These include impacts on infrastructure such as roads, transportation, power grids, water and sanitation systems. Coastal infrastructure, including ports and roads, is susceptible to inundation and coastal erosion from rising sea levels. Intense heatwaves can cause thermal stress such as buckling of roads and railway tracks due to softening and cracking of asphalt, while flooding can damage roads and bridges, disrupting transportation networks. Extreme weather from excessive precipitation and heat can cause power outages impacting essential services. Flooding can damage building foundations and water treatment facilities and can overwhelm sewage systems, leading to contamination of water sources. Prolonged droughts can cause soil contraction and subsidence. Permafrost thaw in high-altitude regions can destabilize foundations for pipelines, roads and railway tracks.

1.4.3 Economic and Distributional Impacts

1.4.3.1 Economic Impacts across Sectors

IPCC estimates that the economic impacts of climate change will be high, especially under higher warming scenarios and at 2 °C warming; annual global GDP losses by mid-century will be between 0.5 and 2.5% IPCC [16]. However, with close to 4 °C, models estimate a 10–23% loss of GDP by 2100. The *Stern Review on the Economics of Climate Change* estimates that the overall costs and risks of climate change will be equivalent to losing at least 5% of global GDP each year, indefinitely and this could rise to 20% of GDP or more [30]. The *Economic Consequences of Climate Change* report [26] estimated that in the absence of further action to tackle climate change, the combined effect on global annual GDP could be between 1.0–3.3% by 2060, which may increase to between 2 and 10% by the end of the century, relative to the no-damage baseline scenario. Although estimates from different studies vary due to differing assumptions and models, there is a consensus that economic damages increase non-linearly with higher warming, with higher impacts on low-income and developing countries.

1.4.3.2 Geographic and Socioeconomic Factors

Climate change has strong distributional impacts and disproportionately affects vulnerable populations. The development stage of the country is an important factor, and low-income countries in the global South face disproportionate risks, despite having contributed the least to historical GHG emissions. Climate vulnerability is non-uniformly distributed, with Least Developed Countries (LDCs), Small Island Developing States (SIDS), and countries in sub-Saharan Africa and South Asia facing higher risk. Countries with high dependence on tourism and agriculture are more vulnerable as they are highly exposed to the impacts of droughts, floods and sea-level rise.

Within countries, socio-economic factors influence vulnerability and people living in informal urban settlements, below the poverty line, and those employed in climate-sensitive sectors such as fishing and tourism face potential loss of livelihoods, especially in SIDS. People in low-income countries have little or no safety nets, and natural calamities like droughts or floods often prove to be financially devastating. Marginalized populations, low-income households, women, the elderly and Indigenous communities often face increased risks due to their limited access to information and relief services, often making them the last beneficiaries of government aid, which exacerbates existing inequalities.

Inequities also exist in adaptive capacity depending on socio-economic status, as low-income countries have limited financial, technological, and institutional capacity to respond to a climate disaster. Post-disaster recovery is slower in low-income countries, while several LDCs depend on international aid, leading to uneven and delayed responses.

1.4.3.3 Cascading and Compounding Risks

Distributional issues and complexities coupled with the interconnectedness of economies and social systems can lead to cascading and compounding risks. This can result in impacts such as climate-induced migration due to sea-level rise, desertification or extreme weather triggering a chain of destabilizing societal events. Economic impacts of loss of livelihood, homes and assets can create increased migration into the mainland. This could potentially change the structural composition of the society, potentially leading to cultural clashes which could result in internal conflicts, especially in regions with existing fragilities. Risks can quickly cascade across sectors and regions and climatic and non-climatic risks can interact and amplify each other. This could potentially paralyze government decision-making, leading to reduced effectiveness of humanitarian response, especially in countries with limited resources and governance capacity.

1.5 Actions for Addressing Climate Change

The Earth's climate is changing, and will continue to do so, with its impact lasting several centuries. Addressing climate change has three interrelated aspects: mitigation, adaptation, and resilience, which act in a fundamentally different way to reduce the underlying causes of climate change, adjust to the impacts of the changing climate, and strengthen the capacity to withstand and recover from climate change, respectively.

Table 1.7 Strategies, actions and examples of climate mitigation

Strategies	Actions	Examples
Energy efficiency	Reducing energy wastage through improved technology	Using efficient cookstoves, LED lamps, better insulation in buildings
Renewable energy	Transition from fossil fuels to renewable energy sources	Scaling up solar, wind, and geothermal power and phasing down oil and gas
Lowering energy demand and circular economy	Lower demand for energy services, reducing, reusing, recycling products	Using public transport and cycling, passive cooling in buildings, design products with alternate material
Fuel switching and electrification	Replacing fossil fuels with low-emission fuels and electricity, use of low-emission hydrogen	Use of sustainable fuels in aviation and shipping, replacing oil-fired boilers with electric boilers
Sustainable agriculture and food system transformation	Incorporating regenerative agricultural practices, sustainable land management, reducing livestock emissions and food waste	Lowering meat intake, use of natural fertilizers, reducing wastage of agricultural produce
Land use and carbon sequestration	Building natural carbon sinks such as forests and wetlands through protection and restoration, using technologies for carbon capture and carbon dioxide removal	Direct air capture, soil carbon enhancement, bioenergy with carbon capture

1.5.1 Climate Change Mitigation

Climate change mitigation refers to actions or activities that limit emissions of GHGs from entering the atmosphere and/or reduce their levels in the atmosphere [19]. Climate mitigation aims to lower GHG emissions but also provides co-benefits such as lower air pollution leading to improved public health and job creation in emerging sectors. Broadly, climate mitigation includes the strategies and actions shown in Table 1.7.

1.5.1.1 Climate Adaptation

Climate adaptation refers to adjustment in ecological, social, or economic systems in response to actual or expected climatic stimuli and their effects or impacts. It includes changes in processes, practices or structures to lower or offset potential damages associated with changes in climate. Adaptation involves adjustments to reduce the vulnerability of communities, regions or activities to changes in climate [12]. Understanding the expected adaptations is essential to assess the impact and

Table 1.8 Strategies, actions and examples of climate adaptation

Strategy	Actions	Examples
Water resource management	Flood risk management, rainwater harvesting, demand-side water conservation	Building desalination plants, reviving natural ponds, rainwater tanks, aquifer recovery, water pricing and governance reforms
Food security	Climate-adaptive crops, sustainable irrigation systems, crop diversification	Planting drought-tolerant crops, drip irrigation systems, local seed banks
Coastal management	Responses to sea-level rise, mangrove restoration, coral reef protection, climate-resilient ports	Building storm surge barriers, mangrove replanting, planned relocation of low-lying habitation in island communities, redundant critical services for port infrastructure
Urban infrastructure	Flood-resilient urban design, climate-proofed transport networks, resilient energy systems	White roofs, urban forests and wetlands, elevated roads and flood-proofed transportation networks, underground power lines, backup microgrids
Ecosystem-based adaptation	Forest conservation, protected area management, watershed management, biodiversity corridors	Rainforest conservation, payment for ecosystem services in tourism, peatland restoration, dedicated wildlife corridors
Health system	Heat-health action plans, disease outbreak surveillance, nutrition security programmes	Heat alerts, malaria early warning systems, community nutrition support
Disaster risk reduction (DRR)	Multi-hazard early warning systems, community-based disaster preparedness, climate risk mapping and modeling	Cyclone early warning system, community disaster drills, national risk assessments, pre-positioned disaster relief material

vulnerability and hence is fundamental to estimating the costs or risks of climate change [12]. As global warming is expected to increase even with a reduction in GHG emissions, adaptation is an important response strategy to climate change which is complementary to climate mitigation. Some important strategies and actions for climate adaptation are shown in Table 1.8.

1.5.1.2 Climate Resilience

Climate resilience is the capacity of interconnected social, economic and ecological systems to cope with, adapt to, and recover from climate hazards, responding or reorganizing in ways that maintain their essential function, identity and structure while also retaining the capacity for transformation [17]. Climate resilience involves

Table 1.9 Strategies, actions and examples of climate resilience

Strategy	Actions	Examples
Community resilience	Local capacity building, social cohesion, indigenous knowledge	Community disaster preparedness committees, village-level vulnerability mapping, community-led watershed management
Infrastructure resilience	Climate-proofing critical utilities and systems (electricity, water, transport), climate risk assessment for new assets	Flood-resilient water treatment plants, backup power and water storage in hospitals, climate risk screening for infrastructure investment
Economic resilience	Diversified livelihoods, parametric insurance mechanisms, social protection system, resilient supply chains	Livelihood diversification, risk insurance facility, social safety net programs, supply chain diversification, coastal blue economy development models
Ecological resilience	Biodiversity conservation, maintaining ecosystem integrity, marine ecosystem management	Rewilding initiatives, marine protection programs, wetland site protection
Institutional resilience	Governance frameworks, adaptive management, climate mainstreaming in planning	National Adaptation Plans (NAPs) and strategy mainstreaming across sectors, iterative climate risk assessment, knowledge updating cycles
Transformative resilience	Systemic change, enabling just transitions, building long-term adaptive capacity	Developing a decarbonized economy, replacing GDP-led growth with alternate metrics

transformation to withstand a harsher environment, which is able to minimize damage and allows long-term functionality in the face of unavoidable climate change. The main climate resilience strategies and actions are shown in Table 1.9.

Effective climate action integrates all three aspects of climate mitigation, adaptation and building climate resilience.

1.6 Conclusion

Climate change has a scientific basis and is backed by evidence of physical changes such as record-breaking GHG concentrations in the atmosphere, global temperatures and sea level rise. Several of the tipping points and planetary boundaries are beyond their safe thresholds, signaling the urgency of immediate action to address climate change.

Despite the strong case to address climate change, global efforts are proving to be inadequate and GHG emissions continue to rise, albeit at a slower rate. The window for climate action is fast closing, and if the current rate of emissions continues, the global carbon budget to limit global warming to 1.5, 1.7 and 2 °C will be exhausted within 4, 12 and 25 years from 2026.

The escalating impacts of climate change are being felt in the form of frequent extreme weather events, increasing sea levels, and ecosystem degradation globally. Physical risks are increasing rapidly and are a threat to food, water and livelihood security. Impacts on natural systems, human systems and economy are growing and the interconnectedness of economies and social systems can lead to compounding risks, especially in countries with limited resources and governance capacity.

Addressing climate change has three interrelated aspects: mitigation, adaptation, and resilience. As there are interconnected challenges, climate solutions must be pursued simultaneously in an integrated manner. This requires action across scales and sectors, which demands coordinated engagement from governments, businesses, civil society, financial institutions, indigenous communities and individuals.

As every fraction of a degree of additional warming translates directly into larger impacts and increases the risk of ecosystem collapse, climate action is no longer a choice but an economic imperative. A climate-resilient future is our only hope and a goal worth pursuing in a climate-constrained world.

References

1. Armstrong McKay DI, Staal A, Abrams JF, Winkelmann R, Sakschewski B, Loriani S, Fetzer I, Cornell SE, Rockström J, Lenton TM (2022) Exceeding 1.5 °C global warming could trigger multiple climate tipping points. Science 377(6611):article eabn7950. https://doi.org/10.1126/science.abn7950. Accessed 20 Jan 2025
2. Climate Action Tracker (no date) Emissions pathways. https://climateactiontracker.org/global/emissions-pathways/. Accessed 20 Jan 2026
3. Climate Watch (2024) Climate watch data. World Resources Institute, Washington, DC. https://www.climatewatchdata.org. Accessed 28 Jan 2025
4. Global Carbon Project (2025) Global carbon budget 2025. https://globalcarbonbudget.org/gcb-2025/. Accessed 30 Jan 2025
5. Global Coral Reef Monitoring Network (2021) Status of coral reefs of the world: 2020—executive summary. https://gcrmn.net/wp-content/uploads/2022/05/Executive-Summary-with-Forewords.pdf. Accessed 28 Jan 2025
6. Greenhouse Gas Protocol (2024) Global warming potential values. https://ghgprotocol.org/sites/default/files/2024-08/Global-Warming-Potential-Values%20%28August%202024%29.pdf. Accessed: 26 Jan 2025
7. Guterres A (2018) Secretary-General's remarks on climate change (as delivered). United Nations. https://www.un.org/sg/en/content/sg/statements/2018-09-10/secretary-generals-remarks-climate-change-delivered. Accessed 10 Jan 2025
8. Intergovernmental Science-Policy Platform on Biodiversity and Ecosystem Services (IPBES) (2019) Summary for policymakers of the global assessment report on biodiversity and ecosystem services. IPBES Secretariat, Bonn. https://files.ipbes.net/ipbes-web-prod-public-files/inline/files/ipbes_global_assessment_report_summary_for_policymakers.pdf. Accessed 12 Jan 2025
9. Intergovernmental Science-Policy Platform on Biodiversity and Ecosystem Services (IPBES) (2019) Global assessment report on biodiversity and ecosystem services. https://www.ipbes.net/global-assessment. Accessed 28 Jan 2025
10. International Institute for Applied Systems Analysis (IIASA) (no date) Shared socioeconomic pathways (SSP). https://iiasa.ac.at/models-tools-data/ssp. Accessed 22 Jan 2025

11. International Institute for Applied Systems Analysis (IIASA) (no date) Representative Concentration Pathways (RCP). https://iiasa.ac.at/models-tools-data/rcp. Accessed 20 Jan 2025
12. IPCC (2001) Climate change 2001: impacts, adaptation and vulnerability—third assessment report, working group II. https://archive.ipcc.ch/ipccreports/tar/wg2/index.php?idp=643. Accessed 29 Jan 2025
13. IPCC (2021) Summary for policymakers. In: Masson-Delmotte V, Zhai P, Pirani A, Connors SL, Péan C, Berger S et al (eds) Climate change 2021: the physical science basis. Contribution of working group I to the sixth assessment report of the intergovernmental panel on climate change. Cambridge University Press, Cambridge and New York, pp 3–32
14. IPCC (2021) Annex VII: glossary. In: Climate change 2021: the physical science basis—working group I contribution to the sixth assessment report. Cambridge University Press, Cambridge. https://www.ipcc.ch/report/ar6/wg1/downloads/report/IPCC_AR6_WGI_AnnexVII.pdf. Accessed 25 Jan 2025
15. IPCC (2022) Climate change 2022: impacts, adaptation and vulnerability—summary for policymakers. https://www.ipcc.ch/report/ar6/wg2/chapter/summary-for-policymakers/. Accessed: 28 Jan 2025
16. IPCC (2022) Climate change 2022: impacts, adaptation and vulnerability—sixth assessment report, working group II. https://www.ipcc.ch/report/ar6/wg2/. Accessed 28 Jan 2025
17. IPCC (2022) Annex II: glossary. In: Climate change 2022: impacts, adaptation and vulnerability. https://www.ipcc.ch/report/ar6/wg2/chapter/annex-ii/. Accessed 29 Jan 2025
18. IPCC (2022) Chapter 2: emissions trends and drivers. In: Climate change 2022: mitigation of climate change. https://www.ipcc.ch/report/ar6/wg3/chapter/chapter-2/. Accessed 26 Jan 2025
19. IPCC (2022) Frequently asked questions—working group III contribution to the sixth assessment report. https://www.ipcc.ch/report/ar6/wg3/downloads/report/IPCC_AR6_WGIII_FAQs_Compiled.pdf. Accessed 29 Jan 2025
20. IPCC (2023) Climate change 2023: synthesis report—summary for policymakers. https://www.ipcc.ch/report/ar6/syr/downloads/report/IPCC_AR6_SYR_SPM.pdf. Accessed 20 Jan 2025
21. IPCC (2023) AR6 synthesis report—headline statements. https://www.ipcc.ch/report/ar6/syr/resources/spm-headline-statements/. Accessed 23 Jan 2025
22. Lenton TM, Armstrong McKay DI, Loriani S, Abrams JF, Lade SJ, Donges JF, Milkoreit M, Powell T, Smith SR, Zimm C, Buxton JE, Bailey E, Laybourn L, Ghadiali A, Dyke JG (eds) (2023) The global tipping points report 2023. University of Exeter, Exeter. https://report-2023.global-tipping-points.org/. Accessed 25 Jan 2025
23. National Aeronautics and Space Administration (NASA) (no date) What is the greenhouse effect? https://science.nasa.gov/climate-change/faq/what-is-the-greenhouse-effect/. Accessed 23 Jan 2025
24. National Oceanic and Atmospheric Administration (NOAA) (2023) Arctic report card 2023: headlines. https://arctic.noaa.gov/report-card/report-card-2023/2023-headlines/. Accessed 28 Jan 2025
25. National Oceanic and Atmospheric Administration (NOAA) (no date) How does climate change affect coral reefs? https://oceanservice.noaa.gov/facts/coralreef-climate.html. Accessed 28 Jan 2025
26. OECD (2015) The economic consequences of climate change. OECD Publishing, Paris. https://www.oecd.org/content/dam/oecd/en/publications/reports/2015/11/the-economic-consequences-of-climate-change_g1g558e1/9789264235410-en.pdf. Accessed 29 Jan 2025
27. Pimm SL, Jenkins CN, Abell R, Brooks TM, Gittleman JL, Joppa LN, Raven PH, Roberts CM, Sexton JO (2014) The biodiversity of species and their rates of extinction, distribution, and protection. Science 344(6187):article 1246752. https://doi.org/10.1126/science.1246752. Accessed 28 Jan 2025
28. Rittel HW, Webber MM (1973) Dilemmas in a general theory of planning. Policy Sci 4(2):155–169. https://doi.org/10.1007/BF01405730. Accessed 13 Jan 2025

29. Romanello M, Walawender M, Hsu S, Moskeland A, Palmeiro-Silva Y, Scamman D et al (2024) The 2024 report of the Lancet countdown on health and climate change: facing record-breaking threats from delayed action. Lancet 404(10465):1847–1896. https://doi.org/10.1016/S0140-6736(24)01822-1. Accessed 12 Jan 2025
30. Stern N (2006) The economics of climate change: the Stern review. Cambridge University Press, Cambridge. https://www.lse.ac.uk/granthaminstitute/publication/the-economics-of-climate-change-the-stern-review/. Accessed 29 Jan 2025
31. Stockholm Resilience Centre (no date) Planetary boundaries. https://www.stockholmresilience.org/research/planetary-boundaries.html. Accessed 25 Jan 2025
32. United Nations (2018) 68% Of the world population projected to live in urban areas by 2050, says UN. https://www.un.org/uk/desa/68-world-population-projected-live-urban-areas-2050-says-un. Accessed 20 Jan 2025
33. United Nations (no date) Global issues: population. https://www.un.org/en/global-issues/population. Accessed 20 Jan 2025
34. United Nations Environment Programme (UNEP) (no date) Projections of future coral bleaching conditions using IPCC CMIP6 models. https://www.unep.org/resources/report/projections-future-coral-bleaching-conditions-using-ipcc-cmip6-models. Accessed 28 Jan 2025
35. United Nations Population Fund (UNFPA) (no date) World population dashboard. https://www.unfpa.org/data/world-population-dashboard. Accessed 20 Jan 2025
36. van Daalen KR, Romanello M, Rocklöv J, Semenza JC, Tonne C, Markandya A, Dasandi N, Jankin S, Achebak H, Ballester J, Bédague D, Blanco-Villafuerte L, Bundele A, Burkart K, Chen B, Chioma Ogudi G, Dasgupta S, Davies M, Di Napoli C et al (2022) The 2022 Europe report of the lancet countdown on health and climate change: towards a climate resilient future. Lancet Reg Health—Europe 22:article 100482. https://doi.org/10.1088/1748-9326/ac7c2a. Accessed 20 Jan 2025
37. World Economic Forum (2025) The global risks report 2025. https://www.weforum.org/publications/global-risks-report-2025/infographics-3e7161574f/. Accessed 15 June 2025
38. World Meteorological Organization (WMO) (2024) Carbon dioxide levels increase by record amount to new highs in 2024. https://wmo.int/news/media-centre/carbon-dioxide-levels-increase-record-amount-new-highs-2024. Accessed 30 Jan 2025
39. World Meteorological Organization (WMO) (2025) State of the Global Climate 2024. WMO, Geneva. https://library.wmo.int/viewer/69455/. Accessed 18 Jan 2025
40. World Meteorological Organization (WMO) (2025) State of the climate 2025—update for COP30. WMO, Geneva. https://wmo.int/sites/default/files/2025-11/State%20of%20the%20Climate%202025%20Update%20COP30%20%2831%20oct%29.pdf. Accessed 10 Jan 2026
41. World Meteorological Organization (WMO) (2025) WMO confirms 2024 as warmest year on record at about 1.55 °C above pre-industrial level. https://wmo.int/news/media-centre/wmo-confirms-2024-warmest-year-record-about-155degc-above-pre-industrial-level. Accessed 23 Jan 2025
42. World Meteorological Organization (WMO) (2026) WMO confirms 2025 was one of warmest years on record. https://wmo.int/news/media-centre/wmo-confirms-2025-was-one-of-warmest-years-record. Accessed 10 June 2026
43. Worldometers (no date) World population by year. https://www.worldometers.info/world-population/world-population-by-year/. Accessed 18 Jan 2025

Chapter 2
Global Climate Governance

Abstract This chapter provides an overview of the global climate governance frameworks and highlights the current challenges. It outlines the global architecture of climate governance with a focus on the existing institutional framework under the Paris Agreement. The chapter also describes complementary approaches in the international maritime and aviation sectors and governance of gases outside of the Paris Agreement. It discusses national and subnational climate governance initiatives and presents the role of non-state actors in climate governance. Lastly, it highlights the challenges and gaps in climate governance before concluding that global climate governance may not be perfect, but the existing frameworks provide flexibility for countries and stakeholders to advance climate action.

Keywords Climate governance · Paris Agreement · Nationally Determined Contributions · Global Stocktake · Enhanced Transparency Framework · Global Climate Action Agenda

2.1 Introduction

Effective and urgent climate action requires strong governance. Global climate governance provides the institutional and accountability frameworks required to translate international climate commitments into real-world action. A well-established and operational global climate governance mechanism ensures that ambitious climate goals and net-zero emissions commitments can be achieved. A streamlined mechanism facilitates international cooperation, improves policy coherence and accelerates systemic transformation in a planned manner while improving transparency of climate action. A flexible and adaptive global climate governance framework is therefore not a choice but an imperative for successful climate action.

This chapter provides a deep dive into global climate governance. Section 2.2 commences with an overview of global climate governance and highlights the current challenges. Section 2.3 outlines the global architecture of climate governance, including the United Nations Framework Convention on Climate Change

K. Narula, *The Climate Catalyst*, Advances in Sustainability Science and Technology,
https://doi.org/10.1007/978-981-92-5213-8_2

(UNFCCC), the Kyoto Protocol and provides a detailed explanation of the Paris Agreement with the existing institutional framework. Section 2.4 describes complementary climate governance in the international maritime and aviation sectors and gases outside of the Paris Agreement. Section 2.5 discusses national and subnational climate governance approaches, including the need for vertical and horizontal integration of climate governance. Section 2.6 presents the role of non-state actors in climate governance along with the international initiatives and frameworks for progressing climate action by corporates. Section 2.7 highlights the challenges and gaps in climate governance before concluding the chapter.

2.2 Overview of Global Climate Governance

2.2.1 Understanding Climate Governance

Climate governance can be understood as the framework of rules, mechanisms, practices, processes, and measures through which governments or the international community can control or influence actions to address climate change. Climate governance is a multilevel and multi-scale process involving several stakeholders, actors and jurisdictions. The governance process is dynamic, continuous and flexible.

Multilevel climate governance is required as addressing climate change entails coordinated efforts across international, supranational, national, regional, and local levels, which involves multiple actors such as governments, businesses, and civil society [53]. Corporate frameworks that address how companies can integrate climate considerations into their decision-making, strategy, and operations also contribute to climate governance. Both formal and informal governance arrangements are important for effective climate governance. Formal arrangements, which include internationally agreed climate agreements, provide the global framework, universal legitimacy, and enforceability in certain cases. They are also important to ensure accountability, rules, procedures and monitoring mechanisms. Informal arrangements can include non-state actor-led initiatives which can influence formal policy-making through advocacy and public engagement. Combining the efforts of governments, corporates and civil society leverages the complementarity of formal and informal approaches in climate governance, leading to more inclusive, equitable, and informed climate action.

2.2.2 Historical Evolution

Early climate scientists included Svante Arrhenius, who in 1896 predicted that a change in atmospheric carbon dioxide would raise the Earth's average temperature through the greenhouse effect. Guy Callendar, in the paper '*The artificial production of carbon dioxide and its influence on temperature*' observed a warming trend and correlated it with carbon dioxide (CO_2) measurements, suggesting that rising atmospheric CO_2 from human activity was responsible for the warming [5]. In the mid-twentieth century, Charles Keeling established the Keeling Curve by continuously monitoring atmospheric CO_2 measurements at the Mauna Loa Observatory, Hawaii. The observation records start at about 315 ppm in 1958 and over time establish a rise in atmospheric CO_2 concentration.

The growing environmental awareness and convergence of scientific consensus in the 1970's was the trigger for establishing a formal process for climate governance. The 1972 Stockholm Conference on the Human Environment was the first major international forum that focused on environmental issues. It highlighted the interconnectedness of human well-being and the environment and established the need for global cooperation for environmental protection.

In 1979, the first World Climate Conference (WCC-1) was held to discuss climate variability, where a declaration for international action to improve understanding of the climate system was adopted. The World Climate Programme (WCP) was established, which laid the foundation for future global climate initiatives [54]. Following this, the United Nations Environment Programme (UNEP), World Meteorological Organization (WMO), and International Council for Science (ICSU) organized the *Villach Conference on greenhouse gases and climate* in 1985, which concluded that increases in GHGs could lead to global warming in the first half of the twenty-first century [24].

The 1988 *Toronto Conference on the Changing Atmosphere* highlighted the global security risks, which politicized the debate. The event brought together over 300 scientists and policymakers from 46 countries and delivered the 'first global scientific consensus' on climate change [55]. It proposed a global target for industrialized nations to reduce CO_2 emissions by 20% from 1988 levels by 2005.

Following the *Toronto Conference,* the WMO and UNEP established the Intergovernmental Panel on Climate Change (IPCC) in 1988. Early assessments by the IPCC included the First Assessment Report (FAR) in 1990, which provided the foundational knowledge base and led to growing scientific understanding of climate change and its impact [56]. This also served as the basis for establishing the 1992 United Nations Framework Convention on Climate Change (UNFCCC). Subsequent reports, including the Second Assessment Report (1996), Third Assessment Report (2001), Fourth Assessment Report (2007), Fifth Assessment Report (2013–2014) and the Sixth Assessment Report (2021–2023) also provided technical insights and evidence on climate change. This established the scientific credibility necessary for informing comprehensive discussions on climate governance.

2.2.3 Governance Challenges

The atmosphere is considered one of the *global commons,* which implies a shared resource and is a *common heritage of mankind*. As it is a shared resource, no single country owns or controls it, but everyone benefits from its use. Countries pursue economic activities that emit GHGs into the shared resource, leading to the degradation of the atmosphere. Hence, while the benefits accrue to individual countries, the costs due to the degraded resource and impacts of climate change are distributed amongst all countries.

Governance of the global commons faces several challenges. These include *the free-rider problem*, where some countries could benefit from the collective efforts of other countries to reduce GHG emissions without contributing to the costs themselves. Due to the different resource endowments and uneven distribution of climate impacts, the costs and benefits of addressing climate change are not perceived equally. Climate governance is also characterized by temporal and spatial misalignments. Countries have competing economic interests and often prioritize short-term economic benefits over long-term consequences of climate change. This has multi-generational impacts. Climate governance also demands acknowledging intergenerational equity and justice, historical responsibility, existing capacity, and future vulnerabilities. For example, vulnerable nations that contributed the least to the problem of GHG emissions often bear a disproportionate cost and impact of climate change, which must be addressed. Challenges include fair burden-sharing and ensuring procedural justice by including marginalized voices in decision-making processes.

Climate governance spans global, national, regional and subnational levels, which creates challenges of coherence and accountability. Fragmentation, duplication, lack of resources, competing development priorities and geopolitical alignments also impact climate governance. Bridging climate science and policy is challenging as there is an inherent uncertainty in climate projections due to climate modeling complexity. Politicization and mistrust exacerbate delays, while information asymmetries hinder decision-making. Climate governance requires application of science, policy, and economics, but differing methods and timeframes prioritize different solutions. Economic models often undervalue future risks and political pressures complicate policy design. Balancing economic growth, environmental impact and social justice makes climate governance inherently difficult, requiring tradeoffs between costs and benefits.

International cooperation across borders is essential to govern the global commons, but as there is no single governing authority, establishing governance frameworks and enforcing international treaties is challenging. The climate governance challenge is compounded by the fact that it operates at a global scale, with nearly 200 nations having different capabilities, interests and priorities, which creates challenges in achieving coordinated action.

2.3 Architecture of Global Climate Governance

Global climate governance has a multifaceted and evolving architecture which extends formally and informally across several entities. Multilateral climate governance involves coordinated international action on climate change through collective agreements and global frameworks. The system has multiple actors, such as international organizations, national governments and non-state entities. The global climate governance landscape is simultaneously centralized and decentralized, with overlapping institutions that shape action across sectors. Climate governance is progressed through both binding agreements and flexible approaches that respond to evolving scientific and political realities to collectively steer global climate action.

2.3.1 The United Nations Framework Convention on Climate Change

The United Nations Framework Convention on Climate Change (UNFCCC) is the parent treaty for international climate action and provides a framework for global climate governance. The UN General Assembly Resolution 45/212, adopted in December 1990, established the Intergovernmental Negotiating Committee for a Framework Convention on Climate Change to negotiate an international treaty on climate change. This resolution was the starting point for the formal diplomatic process that led to the creation of the UNFCCC.

The UNFCCC was formally created at the United Nations Conference on Environment and Development (UNCED) held in Rio de Janeiro in 1992 as an international treaty to address climate change and came into force in 1994. It provided the foundation for a collaborative global approach to address climate change, including consensus-based decision-making.

The Convention provides the overarching legal and political framework for addressing climate change. The principle of *common but differentiated responsibilities and respective capabilities* (CBDR-RC) is one of the core principles of the UNFCCC to achieve equity in international climate policy. This implies that all countries have a shared responsibility to address climate change. However, it recognizes that all countries are not equally responsible for causing climate change and have different capacities to address the issue. Further, developing countries have the right to pursue economic growth and social development and in accordance with the principle of equitable development, their commitments must be balanced with these needs. Therefore, different countries have different obligations, with developed countries bearing the majority responsibility for climate action.

The *precautionary principle* is embedded in the Convention, which guides decision-making under scientific uncertainty, allowing for proactive climate action. Sustainable development is the cornerstone that guides climate action to support

Table 2.1 Key features of the Kyoto Protocol

Key features	Details
Binding targets	Legally binding targets to reduce GHG emissions by an average of 5% compared to the 1990 level
Involved parties	Binding targets for 37 industrialized (Annex I) countries and the European Union (EU). Developing (non-Annex I) countries were not required to take action but could participate voluntarily in market mechanisms[1]
Ratification	Entered into force in 2005 after ratification by 55 countries and eventually ratified by 192 parties (191 states and the EU)
Commitment period	• First: 2008–2012 • Second: 2013–2020 (via the Doha amendment)
Approved market mechanisms	• International emissions trading • Clean development mechanism (CDM) • Joint implementation (JI)
GHGs covered	• CO_2, CH_4, N_2O, HFCs, PFCs, SF_6. • NF_3 was added in 2012 and was applicable for the second commitment period

economic, environmental and social progress with an emphasis on equity and fairness. The Convention relies on progress through promoting international cooperation via mechanisms for technology transfer, financial assistance, and information sharing. The Convention established the formal process of climate governance through creating an annual Conference of the Parties (COP) to assess progress in addressing climate change. The Kyoto Protocol and the Paris Agreement operationalize the UNFCCC.

2.3.2 *The Kyoto Protocol*

Institutionalization of climate governance under the UNFCCC led to the Kyoto Protocol, which was the first international climate treaty that legally required developed countries to reduce their GHG emissions. It was adopted in 1997 and played an important role in the acceptance of the legal and institutional frameworks for international climate governance. Its key features are summarized in Table 2.1.

The Kyoto Protocol was a top-down regulatory approach to reduce GHG emissions. It marked a major shift by imposing legally binding emission reduction targets on industrialized nations, in contrast to the earlier voluntary approaches. It introduced

[1] There were different targets for different countries. The EU had an overall target to reduce its GHG emissions by 8% below 1990 levels, with individual targets set for Member States through a burden-sharing agreement.

internationally negotiated Quantified Emission Limitation and Reduction Commitments (QELRCs) to track each country's progress and compliance with its emissions reduction targets. These were subject to international law and had legal consequences for non-compliance. This framework allowed ease of measurement and created accountability through standardized methodologies with international oversight, creating a compliance-oriented system for climate action. Among the major economies, the United States signed the protocol but did not ratify it, while Canada withdrew from the protocol in December 2012, before the end of the first commitment period.

The second commitment period (2013–2020) was established through the Doha Amendment in 2012, which strengthened GHG reduction commitments by setting an aggregate reduction target of at least 18% below 1990 levels for 37 countries and the EU. The Amendment entered into force on 31 December 2020, when it was ratified by the mandated minimum of at least 144 states, which was also the last day of the second commitment period. Several major emitters like Japan, New Zealand and Russia chose not to take on new binding targets in the second commitment period.

To reduce compliance costs, the Kyoto Protocol introduced three market-based flexibility mechanisms—international emissions trading, the Clean Development Mechanism (CDM) and Joint Implementation (JI). These mechanisms collectively created the foundation for international carbon markets and allowed international funding for undertaking projects in countries with lower GHG abatement costs. International emissions trading allowed Annex I countries to buy and sell emission reduction credits to meet their GHG reduction targets by purchasing excess allowances from other countries. CDM allowed developed countries to invest in emission-reduction projects in developing countries to earn tradable Certified Emission Reduction (CER) credits, which could be used toward their compliance obligations against their own GHG reduction targets. JI allowed an Annex I country to invest in an emission-reduction project in another Annex I country to generate Emission Reduction Units (ERUs) through collaborative efforts which could be counted toward its own GHG reduction commitment. However, there were concerns about environmental integrity, including additionality in CDM projects, price volatility of CERs, and issues related to project validation and monitoring.

The Kyoto Protocol established a robust compliance system as an integral part of the protocol. It has a compliance committee with a facilitative and enforcement branch to promote adherence to commitments and address non-compliance cases [36]. The facilitative branch provides advice and assistance to Parties for promoting compliance, while the enforcement branch could impose consequences on Annex I Parties that failed to meet their commitments, including suspension from the flexibility mechanisms and penalties. This compliance architecture was a milestone in international climate governance as it introduced consequences for non-compliance, beyond relying on international diplomatic pressure and naming-and-shaming.

The Kyoto Protocol highlighted the political challenges of implementing a top-down climate governance architecture but also demonstrated the power of binding international climate law. While the Kyoto Protocol's effectiveness remains contested, its overall impact was significant as it established a framework for international climate governance, demonstrated the feasibility of international carbon markets, raised global awareness and catalyzed international efforts to address climate change.

2.3.3 *The Paris Agreement*

Non-inclusion of high-emitting developing countries in the Kyoto Protocol led to a search for a new model of climate governance. The top-down approach with binding commitments only for Annex 1 countries was challenged in the face of growing global economic competitiveness and rapidly increasing GHG emissions from emerging economies. Formal negotiations continued at the annual meeting of the UNFCCC but came to a standstill with disagreements over new binding targets. This ultimately gave way to a new bottom-up model of climate governance that addressed the need for larger participation, the reality of non-binding targets and providing flexibility in climate action.

The Paris Agreement is a legally binding international treaty on climate change where the processes (and not commitments) are legally binding. It was adopted at the UN Climate Change Conference (COP21) in Paris, France, on 12 December 2015 by 195 Parties and entered into force on 4 November 2016 [35]. It contains 29 Articles that establish a framework for addressing climate change, including mitigation, adaptation and loss and damage.

The overarching goal of the Paris Agreement is to hold 'the increase in the global average temperature to well below 2 °C above pre-industrial levels' and pursue efforts 'to limit the temperature increase to 1.5 °C above pre-industrial levels' [37]. The Agreement acknowledges that this requires global GHG emissions to peak as soon as possible and to reach net-zero emissions in the second half of this century. The Paris Agreement also seeks to strengthen the ability of countries to adapt to the adverse impacts of climate change and build climate resilience. It enables voluntary cooperative approaches through facilitating carbon markets, and addresses loss and damage from climate change.

The Paris Agreement establishes a system of support for developing countries, focusing on finance, technology, and capacity-building. Developed countries are expected to provide and mobilize financial resources to fund mitigation and adaptation action in developing countries, and other Parties are encouraged to provide voluntary support. An established framework for technology development guides the transfer of climate-safe technologies and calls for supporting enhanced capacity building for effective climate action in developing countries.

The Paris Agreement pivoted from being limited to a few countries to near-universal participation, making all countries willing participants in collective climate governance. The flexible approach respected the differing national circumstances and provided countries with more autonomy over their climate policies, which led to political consensus over the agreement. The Paris Agreement was successful in replacing the Kyoto Protocol as it shifted from a top-down, punitive system with limited participation to a bottom-up, flexible, and universal framework. It has become a landmark in the multilateral climate governance process, which brings all countries together to combat climate change and adapt to its effects. The key elements of the Paris Agreement framework include the following.

2.3.3.1 Nationally Determined Contributions

Nationally Determined Contributions (NDCs) embody efforts by each country to reduce national emissions and adapt to the impacts of climate change. The framework is designed to drive ambition through adoption of progressively stronger commitments. Each participating country sets its own GHG reduction targets. It is not legally binding to achieve the set targets, but countries are obligated to submit them, pursue domestic actions to fulfill their commitments and report on their progress. The success of the Paris Agreement is eventually dependent on the collective and effective implementation of self-determined NDCs. Regular reporting on progress towards meeting targets fosters transparency and accountability.

The NDCs follow a defined format as countries must provide information to ensure clarity and transparency of target setting and reporting under the Paris Agreement. The narrative document contains a comprehensive description of the country's climate plan, including the implementation period for the NDC; sectors and GHGs covered; details on national adaptation strategies, plans, and actions; mitigation implementation plans and strategies; description of the domestic processes and stakeholder consultations for preparing the NDC along with other supporting details.

Parties are also required to submit information in a common tabular format, which are standardized tables used to report and track progress. The predefined format ensures that the information is provided in a quantified, consistent, comparable and transparent manner. It includes information on the base year(s) for the NDC targets; quantifiable emission reduction goals including target year and baseline emission levels; metrics used to track progress toward the implementation and achievement of the NDC; GHG emissions accounting methodologies and assumptions; details on financial, technical and capacity-building support provided and received, and other details.

2.3.3.2 The Ratchet Mechanism and Increasing Ambition

The Paris Agreement incorporates a five-year cycle and mandates countries to submit the updated NDCs every five years. Each new NDC is expected to increase the commitment and reflect the country's 'highest possible ambition' in an effort to collectively reach the long-term goals of the Paris Agreement. This ensures that commitments can be continually strengthened and updated to reflect the latest climate science, technological developments and respective capabilities.

Before the Paris Agreement was formally adopted, countries submitted their initial climate action plans, known as Intended Nationally Determined Contributions (INDCs), throughout 2015, which then became NDC 1.0 for parties that ratified the agreement. The second round of NDCs was due in 2020 with targets for 2030, but there were several delays and late submissions due to the COVID-19 pandemic. NDCs 3.0 were due in February 2025 but only 13 Parties had submitted them by the February 10, 2025, deadline. By 31 December 2025, 132 Parties had submitted their NDCs outlining their climate actions through 2035 [38].

2.3.3.3 Global Stocktake

The Paris Agreement mandates a Global Stocktake (GST) every five years, starting in 2023, to assess the collective progress made towards achieving the goals and to identify the gaps for further action. The first GST took place at COP28 and confirmed that the world is off track to meet the goals of the Paris Agreement. It also acknowledged the need to cut global GHG emissions by 43% by 2030 (compared to 2019 levels) to limit global warming to 1.5 °C. This synthesis provided feedback to inform countries in preparing their NDC 3.0. Other main outcomes included a call for urgent action to accelerate a just transition away from fossil fuels, tripling renewables and doubling energy efficiency by 2030; strengthening adaptation, including scaling up adaptation finance and delivery of National Adaptation Plans (NAPs); and the operationalization of the Loss and Damage Fund. The process for the second GST to evaluate collective progress on climate action to inform future climate pledges will begin in late 2026, and is due to conclude in 2028.

2.3.3.4 Enhanced Transparency Framework

The Enhanced Transparency Framework (ETF) is a central mechanism for a universal reporting system to build trust, promote accountability, and track global progress under the Paris Agreement. It requires all signatory countries to report their climate actions and progress through submission of the Biennial Transparency Reports (BTRs) every two years. BTRs contain GHG inventories, progress on NDCs, adaptation actions, information on financial, technological, and capacity-building support provided and mobilized by developed countries and needed and received by developing countries. The first BTR submission date was December 31, 2024, followed by the second BTR submission in December 31, 2026.

The ETF provides a universal framework but includes built-in flexibility for developing countries, particularly Least Developed Countries (LDCs) and Small Island Developing States (SIDS). Countries can choose to use this flexibility based on their national capacities while stating their constraints and need to provide a timeframe for future improvements in their reporting. The ETF outlines a rigorous two-step international review and evaluation process to ensure that the reported information is transparent, accurate, and consistent. A team of technical experts conducts a Technical Expert Review (TER) of each country's BTR to check for adherence to the reporting framework and to identify areas for improvement. The data from BTRs is used as an input for the GST to assess the collective progress towards the Paris Agreement's long-term goals and is critical to its five-year ambition cycle.

The Facilitative, Multilateral Consideration of Progress (FMCP) is a key component of the ETF for all parties where countries engage in an open, multilateral dialogue to review and discuss their climate actions. It encourages transparency and learning, builds collective understanding through fostering dialogue and allows developing countries to highlight their capacity-building needs for climate action and reporting.

2.3.3.5 Long-Term Low GHG Emission Development Strategies

Long-term low greenhouse gas emission development strategies (LT-LEDS) are comprehensive national roadmaps that outline how countries plan to transition towards a low-carbon, climate-resilient economy by 2050. These are not mandatory but encouraged under the Paris Agreement and provide a long-term vision that complements the shorter-term NDCs. These help to align near-term climate actions and investments with broader, sustainable development goals, offering guidance and direction for long-term planning and decision-making. LT-LEDS include a long-term vision, detailed sectoral pathways to achieve them, links between the LT-LEDS, NDCs and a country's national development plans. It also includes an assessment of the financial investments required, showcases stakeholder engagement, and provides a framework for tracking progress.

2.3.4 The Institutional Framework

The UNFCCC operates through a comprehensive institutional framework to support international climate cooperation, which includes the following.

(a) Governing Bodies

There are three governing bodies. The Conference of the Parties (COP) is the main decision-making body of the Convention [39]. The Conference of the Parties serving as the meeting of the Parties to the Kyoto Protocol (CMP) oversees the implementation of the Kyoto Protocol [40]. The Conference of the Parties serving as the Meeting of the Parties to the Paris Agreement (CMA) is responsible for overseeing the implementation of the Paris Agreement, and making decisions to promote its effective enforcement [41]. Ad hoc subsidiary bodies can also be established as needed to address specific issues.

(b) Permanent Subsidiary Bodies

These consist of the Subsidiary Body for Scientific and Technological Advice (SBSTA) which supports the work of the COP, the CMP and the CMA through the provision of timely information and advice on scientific and technological matters as they relate to the Convention, its Kyoto Protocol and the Paris Agreement [42] The Subsidiary Body for Implementation (SBI) is responsible for implementation under the Convention, the Kyoto Protocol and the Paris Agreement and aims at enhancing the ambition of Parties [43]. Its agenda is shaped around the key building blocks of implementation of treaties and instruments, including transparency, mitigation, adaptation, finance, technology and capacity building.

(c) United Nations Climate Change

The United Nations Climate Change (formally known as the UNFCCC secretariat) was established in 1992. This is the permanent secretariat that provides organizational support and technical expertise and enables the Convention to function effectively through its administrative and technical work.

The UNFCCC operates through a robust decision-making process to address climate change. Meetings involve negotiations and debates which ensure inclusivity, transparency, and respect for the national circumstances of all Parties. As decisions are taken by consensus and not by majority voting, all Parties must agree to adopt a resolution, which sometimes delays decision-making and may result in slow progress.

2.3.5 UNFCCC COP Outcomes

The UNFCCC COP meetings can be characterized into three phases of climate governance. The first phase from 1995 to 2001 was used to build a framework for binding targets. In this phase, COP3 (Kyoto), where top-down binding targets were agreed, was the defining moment. The second phase from 2007 to 2015 can be considered the negotiation and transition phase. COP15 (Copenhagen), where a 2° target was introduced, and COP21 (Paris), where a bottom-up architecture was agreed, were the most transformative COPs in this era. The third phase from 2016 onwards can be considered the implementation and ambition era. COP24 (Katowice), where the Paris rulebook was adopted, and COP28 (Dubai), where the first GST was undertaken, are important milestones in this phase. The chronological list of the COP meetings and the key outcomes which have shaped global climate governance is shown in Table 2.2.

Table 2.2 UNFCCC COP meetings and key outcomes

COP	Year	Location	Key outcomes
COP1	1995	Berlin, Germany	Negotiations commence
COP2	1996	Geneva, Switzerland	IPCC report endorsed as the basis for climate action
COP3	1997	Kyoto, Japan	Kyoto protocol adopted
COP4	1998	Buenos Aires, Argentina	Kyoto mechanisms operationalized
COP5	1999	Bonn, Germany	Negotiations on implementing Kyoto protocol
COP6	2000	The Hague, Netherlands	Talks end in deadlock over disagreements
COP6-2	2001	Bonn, Germany	Revived stalled Kyoto negotiations from COP6
COP7	2001	Marrakesh, Morocco	Kyoto rulebook for market mechanisms finalized
COP8	2002	New Delhi, India	Sustainable development and adaptation focus
COP9	2003	Milan, Italy	Operational guidelines for CDM finalized
COP10	2004	Buenos Aires, Argentina	Discussions start on post-2012 climate regime
COP11	2005	Montreal, Canada	Kyoto protocol enters into force
COP12	2006	Nairobi, Kenya	Nairobi work programme on climate impacts and adaptation established
COP13	2007	Bali, Indonesia	Negotiations commence toward a post-Kyoto climate framework
COP14	2008	Poznan, Poland	Discussions on adaptation funding
COP15	2009	Copenhagen, Denmark	Voluntary national pledges and 2 °C goal introduced
COP16	2010	Cancun, Mexico	Mitigation pledges formalized, establishment of the green climate fund
COP17	2011	Durban, South Africa	Process commences leading to the Paris agreement
COP18	2012	Doha, Qatar	Kyoto protocol extended to 2020
COP19	2013	Warsaw, Poland	Warsaw international mechanism for loss and damage established
COP20	2014	Lima, Peru	Framework for INDCs introduced
COP21	2015	Paris, France	Paris agreement adopted
COP22	2016	Marrakesh, Morocco	Paris agreement implementation discussions
COP23	2017	Bonn, Germany (Fiji presidency)	Advanced Paris rulebook
COP24	2018	Katowice, Poland	Adoption of the Paris rulebook

(continued)

Table 2.2 (continued)

COP	Year	Location	Key outcomes
COP25	2019	Madrid, Spain (Chile presidency)	Negotiations on article 6 carbon markets fail. Focus on oceans
COP26	2021	Glasgow, United Kingdom	Article 6 finalized, coal phase-down introduced
COP27	2022	Sharm El-sheikh, Egypt	Agreed to establish a loss and damage fund
COP28	2023	Dubai, UAE	First GST conducted; adopted call to transition away from fossil fuels in energy system
COP29	2024	Baku, Azerbaijan	Set a new collective quantified goal (NCQG) for global climate finance
COP30	2025	Belém, Brazil	Focus on enhanced NDCs

2.4 Complimentary Climate Governance

2.4.1 Sectoral Approach for International Shipping and Aviation

GHG emissions from international shipping and aviation are considered outside the definition of national emissions, as these do not happen within the borders of a country. This made emissions from international shipping and aviation problematic to regulate under the UNFCCC framework, which was designed for national emissions.

Traditionally, bunkers, which deliver fuels to merchant ships and civil aircraft undergoing international voyages, are not included in domestic consumption or as exports in energy accounting. Further, there was no consensus on the process of allocating emissions from these fuels to individual countries. For example, it could not be agreed if GHG emissions from these sectors should be included in the country where the journey started or ended, where the aircraft or the ship is registered, or where the fuel was sold. As Parties could not agree on which country should include GHG emissions from bunker fuels, it was decided to club them together as emissions for international services and operations.

Due to the methodological and legal challenges associated with accounting of GHG emissions from international aviation and shipping, it was agreed under Article 2.2 of the Kyoto Protocol that Annex I Parties shall pursue reduction of these GHG emissions working through the International Civil Aviation Organization (ICAO) and the International Maritime Organization (IMO) [52]. This sectoral approach was continued in the Paris Agreement, and while GHG emissions from domestic aviation and shipping are included in national targets for countries, international shipping and aviation continue to be governed through IMO and ICAO, which regularly exchange emission information with the UNFCCC [18].

2.4.1.1 IMO Led Framework for International Shipping

In 2022, international shipping accounted for about 2% of global energy-related CO_2 emissions, which have been rising continuously [21]. The IMO governs international shipping regulations and has addressed GHG emissions through a series of mandatory measures since the 1990s. In September 1997, the IMO adopted Resolution 8 to address CO_2 emissions from ships. This initiated the IMO's work on GHG emissions, which included a study on CO_2 emissions, establishing a precedent for future studies. In 2011, mandatory technical and operational measures for ships were introduced under Annex VI of the MARPOL Convention (International Convention for the Prevention of Pollution from Ships) by the IMO for certain ships. This included the Energy Efficiency Design Index (EEDI), which required new ships to meet minimum energy efficiency standards, and the Ship Energy Efficiency Management Plan (SEEMP), which required ships to have an operational plan for improving energy efficiency. This was followed by adoption of a mandatory global Data Collection System (DCS) in 2016, which required ships to report their fuel oil consumption and laid the foundation for subsequent measures.

The *Initial IMO Strategy on reduction of GHG emissions from ships* (2018) was the first major international framework to address GHG emissions from international shipping with a long-term vision to phase out shipping's GHG emissions as soon as possible. Total annual GHG emissions from international shipping were to be reduced by at least 50% by 2050, as compared to the 2008 level. It also aimed to decrease carbon intensity (CO_2 emissions per transport work) by at least 40% by 2030, while pursuing efforts towards a 70% reduction by 2050. The strategy acknowledged the need for shipping to contribute to the Paris Agreement's climate goals and emphasized that the impacts of these measures on SIDS and LDCs must be considered.

In 2021, the IMO adopted mandatory technical and operational measures to improve energy efficiency and lower the carbon intensity for existing ships, which came into force on January 1, 2023. These included the Energy Efficiency Existing Ship Index (EEXI) and the Carbon Intensity Indicator (CII) under MARPOL Annex VI. In July 2023, the IMO adopted the *2023 IMO Strategy on Reduction of GHG Emissions from Ships*, which provided a framework for decarbonizing international shipping with a long-term goal of achieving net-zero GHG emissions by or around 2050. It aimed to reduce total annual GHG emissions by at least 20% (striving for 30%) by 2030, and by at least 70% (striving for 80%) by 2040. It also aims to increase the uptake of zero or near-zero GHG emission fuels and technologies to at least 5% (preferably 10%) of the energy used by international shipping by 2030.

The *IMO Net-Zero Framework* was approved by the Marine Environment Protection Committee (MEPC) in April 2025. It is aimed at decarbonizing the international shipping industry by or around 2050. The legally binding agreement was expected to be adopted in October 2025, but discussions have been adjourned until 2026. 16 months after the adoption, the regulations are expected to enter into force, following which governments will be responsible for enforcing them.

The regulations, once approved, will be applicable to all large ocean-going ships of 5000 gross t and above engaged in international trade, which account for approximately 85% of the sector's total CO_2 emissions. The framework proposes to introduce a global marine fuel standard with targets to reduce the annual GHG fuel intensity (GFI) of a ship over time. It also proposes a GHG emissions pricing mechanism which includes a market-based measure to create financial incentives and penalties for GHG emissions. An IMO Net-Zero Fund would collect penalties from non-compliant ships, which will be used to support initiatives for decarbonization of shipping and capacity building to support a just and equitable transition [19].

2.4.1.2 ICAO-Led Mechanism for International Aviation

GHG emissions from international aviation have been growing rapidly in the last two decades. In 2023, they accounted for 2.5% of global energy-related CO_2 emissions [22]. As the Kyoto Protocol and the Paris Agreement did not include the international aviation sector, the ICAO adopted the Carbon Offsetting and Reduction Scheme for International Aviation (CORSIA) in 2016 to address CO_2 emissions from international aviation. This applies to all 193 member States of the ICAO from January 01, 2019 [20].

CORSIA is a global market-based measure to cap and offset emissions from international flights. It is historic as it was the first instance where an entire industry agreed to a global market-based measure to address climate change. The CORSIA works through two main mechanisms.

(a) Carbon offsetting, where airlines must purchase and cancel CORSIA Eligible Emissions Units (CEEUs), which are approved carbon credits sourced from emissions reduction projects that meet strict eligibility criteria set by ICAO and adhere to well-accepted carbon offset certification standards [14].
(b) Use of alternative fuels such as Sustainable Aviation Fuel (SAF) and Lower Carbon Aviation Fuel (LCAF), collectively known as CORSIA Eligible Fuels (CEF).

CORSIA adopted three distinct phases for implementation. The pilot phase from 2021 to 2023 was voluntary and focused on monitoring, reporting, and verification (MRV) of emissions, where only the participating states were subject to offsetting requirements. The baseline for emissions was set at the 2019 level for the first phase. Due to the impact of the COVID-19 pandemic, there was a significant drop in international travel during 2021–2023, and emissions were well below the 2019 baseline, which implied that no offsetting was required by airlines.

In the first phase from 2024 to 2026, the baseline was changed to 85% of 2019 emissions, and it is most likely that this phase will require offsetting. Participation is still voluntary for countries, and offsetting obligations apply to flights between all participating states. The mandatory phase begins from 2027, and all international flights will be subject to offsetting requirements. However, flights to and from LDCs, SIDS, Landlocked Developing Countries (LLDCs) and States that represented less than 0.5% of global international Revenue Tonne Kilometers (RTK) in 2018 will be exempt from offsetting requirements, unless these States participate voluntarily [20].

In 2022, the ICAO adopted the long-term global aspirational goal to achieve net-zero carbon emissions by 2050 from international aviation in support of the Paris Agreement's temperature goal Organization [15]. As a part of this goal, ICAO is pursuing a basket of measures including technological innovations such as deploying electric and hydrogen-powered aircraft, improvements in aircraft design, operations, and air traffic management to reduce fuel consumption and emissions.

The ICAO has also set a collective global aspirational vision for international aviation to be 5% less carbon-intensive by 2030 through the use of SAF. This target was agreed at the Third Conference on Aviation Alternative Fuels (CAAF/3), held in Dubai in November 2023, by over 100 countries. However, the 5% reduction goal is voluntary and not a binding commitment for individual countries. Several countries like Norway, Sweden, France, and the United Kingdom have adopted SAF mandates with targets ranging from 0.5 to 2%. The EU mandates a 2% SAF share under the ReFuelEU Aviation initiative commencing from 2025. This imposes legally binding targets for SAF blending into conventional jet fuel for all EU airports (with >800,000 passengers or >100,000 tonnes of freight annually). This mandate is proposed to be increased to 6% in 2030 and up to 70% by 2050. Thailand and Singapore also adopted a 1% SAF blending mandate for 2026, and other Asian countries are also planning to introduce these mandates in the next few years.

2.4.2 *Governance of Other GHGs*

Seven GHGs were part of the expanded Kyoto Protocol, and the same basket has been retained for the Paris Agreement. These include Carbon Dioxide (CO_2), Methane (CH_4), Nitrous Oxide (N_2O), Hydrofluorocarbons (HFCs), Perfluorocarbons (PFCs), Sulphur Hexafluoride (SF_6), and Nitrogen Trifluoride (NF_3) (added in 2012 through the Doha Amendment). However, there are other gases which contribute to global warming but do not fall under the existing Paris Agreement. These are governed through other complementary mechanisms.

2.4.2.1 Montreal Protocol and Kigali Amendment

The *Vienna Convention for the Protection of the Ozone Layer*, adopted in 1985, is a global treaty to protect the ozone layer. It is a framework convention for international cooperation to promote scientific research, exchange information and monitoring on ozone depletion, but does not require countries to take any actions [26]. The Convention led to the legally binding *Montreal Protocol on Substances that Deplete the Ozone Layer* (commonly known as the *Montreal Protocol*), which was adopted in 1987 and entered into force in 1989. This is an agreement to phase out the production and consumption of ozone-depleting substances (ODS) such as Chlorofluorocarbons (CFCs) and Halons. Along with the Vienna Convention, it achieved universal participation in September 2009.

The main aim to phase out ODS is to protect the stratospheric ozone layer, which stops harmful ultraviolet radiations to reach the Earth. The *Montreal Protocol* covers several chemicals which were used in refrigeration, air conditioning, foam production, and fire extinguishing systems. These chemicals persisted in the atmosphere for decades, releasing chlorine and bromine atoms that destroyed the ozone layer. Some of these ODS have GWPs that are thousands of times higher than CO_2, which can create significant warming and include the following [27].

Annex A Chlorofluorocarbons (e.g., CFC-11, CFC-12) and Halons (e.g., Halon-1301, Halon-1211).

Annex B Other substances such as other fully halogenated CFCs, Carbon Tetrachloride (CTC) and 1,1,1-Trichloroethane (Methyl Chloroform).

Annex C Hydrochlorofluorocarbons (HCFCs), Hydrobromofluorocarbons (HBFCs) and Bromochloromethane (BCM).

Annex E Methyl Bromide (MB) used in fumigation.

Annex F Hydrofluorocarbons (HFCs). These were substitutes for ODS like CFCs and HCFCs and have negligible ozone depletion potential. These were added due to their high GWP.

The *Kigali Amendment* is an international agreement, adopted in 2016 under the *Montreal Protocol*, that extends its scope to include HFCs. The amendment aims at a gradual reduction with a global phase-down of HFCs by 80–85% by 2047, which is expected to prevent a global temperature increase of 0.3–0.5 °C by the end of the century. The inclusion of HFCs signifies the importance of collectively addressing climate governance and the interconnected nature of multilateral environmental agreements.

The *Montreal Protocol* is considered the most successful global environmental agreement in the history of the United Nations, as it has phased out 98–99% of ODS since its adoption, which has led to the notable recovery of the ozone layer [31]. Although it was created to protect the ozone layer, the phasing out of ODS has avoided large amounts of global warming, making it a major climate mitigation framework.

2.4.2.2 Short-Lived Climate Pollutants

Short-lived climate pollutants (SLCPs) are potent warming agents that stay in the atmosphere for a shorter duration than carbon dioxide but have much higher warming potential. Many SLCPs, including black carbon, tropospheric ozone, methane, and HFCs, contribute up to 45% of current global warming and could account for half of human-caused warming in the coming decades if emissions are not reduced [7].

Black carbon, commonly known as soot, is a component of fine particulate air pollution ($PM_{2.5}$) and is released during the incomplete combustion of fuels such as wood, charcoal, and waste. The average atmospheric lifetime of black carbon particles is 4–12 days, but it has a warming impact up to 1500 times stronger than CO_2 per unit of mass, which makes it a potent climate forcer [8]. Tropospheric ozone (O_3) is formed by the interaction of sunlight with methane, volatile organic compounds (VOCs) and nitrogen oxides (NO_X), which are emitted from vehicles and industry. It has an atmospheric lifetime ranging from a few hours to a few weeks and is a super pollutant which is responsible for approximately 0.23 °C of present-day global warming [6]. Methane is emitted from the extraction of fossil fuels, agriculture, livestock and waste. It is a potent SLCP which has contributed to almost a third of net warming since the Industrial Revolution [12]. As short-lived climate pollutants remain in the atmosphere for short time periods ranging from hours to 15 years, reducing their emissions can play an important role in slowing the rate of global warming. It is estimated that maximum possible reductions in SLCPs can achieve a 0.6 °C reduction in warming by 2050, making it a strong contributor to climate action along with the Paris Agreement [7].

The governance of SLCPs is fragmented and uneven. While methane is covered under the list of GHGs governed under the Paris Agreement and HFCs are included in the Montreal Protocol, governance of black carbon and tropospheric (or ground-level) ozone is weak and relies on voluntary action. Overall, SLCPs are still managed through overlapping instruments rather than a single, overarching, legally binding global treaty.

Global climate governance has increasingly recognized the importance of addressing SLCPs in conjunction with a broader framework of climate governance. The Climate and Clean Air Coalition (CCAC), founded in 2012, is a voluntary partnership of over 200 governments, intergovernmental organizations, and non-governmental organizations, and is convened within UNEP. It works to reduce SLCPs that drive both climate change and air pollution with targeted mitigation action within countries and sectors [9]. The CCAC also supports the Global Methane Pledge (GMP) launched on the sidelines of COP26 in 2021. It is a voluntary framework to collectively reduce methane emissions by 30% (from the 2020 level) by 2030, which could eliminate over 0.2 °C of warming by 2050. The GMP has over 155 country participants, representing about half of the global anthropogenic methane emissions. Assessment shows that human-caused methane emissions can be cost-effectively reduced by up to 45% this decade with additional co-benefits such as air quality improvement and health benefits. Although SLCP governance is not perfect, efforts are underway to harmonize actions such that they complement long-term climate goals under the Paris Agreement.

2.5 National and Subnational Climate Governance

2.5.1 National Frameworks and Instruments

National climate governance frameworks have evolved over the years, and countries have adopted different models and structures. After the Paris Agreement in 2015, several countries have established comprehensive institutional frameworks for climate governance, which are backed by legal and regulatory mechanisms. The institutional architecture of national climate governance is often composed of executive bodies, legal and policy frameworks, specialized agencies/institutions tasked with implementation, supported by scientific advisory bodies and dedicated monitoring agencies.

Several countries have established climate ministries and specialized agencies and institutions for implementing and overseeing national climate policy. These agencies have the responsibility of translating international climate commitments into national action plans and implementing coordinated policies across different sectors. The climate change portfolio is mainly a part of the environment ministry. Some examples include India which has the Ministry of Environment, Forest and Climate Change (MoEFCC); the United Kingdom having the Department for Energy Security and Net Zero (DESNZ); Canada where the Minister of Environment and Climate Change leads the department of Environment and Climate Change Canada (ECCC); Fiji having the Ministry of Environment and Climate Change; United Arab Emirates having the Ministry of Climate Change and Environment (MOCCAE); Brazil with the Ministry of Environment and Climate Change (MMA); and Australia having the Climate Change Authority (CCA).

Key climate planning instruments at the national level include Nationally Determined Contributions (NDCs), which set overall mitigation and adaptation commitments. Long-Term Low Emission Development Strategies (LT-LEDS) are voluntary, longer-term strategies to address climate change while pursuing climate-resilient development by 2050. Till the end of 2025, 80 countries had submitted their LT-LEDS [44]. National Adaptation Plans (NAPs) provide strategies to address adaptation needs, particularly for developing countries and LDCs to create medium- to long-term strategies for adaptation. It allows countries to assess their vulnerabilities, reduce them by building resilience and to integrate climate change adaptation into national development planning [45]. About 75 countries had submitted their NAPs by the end of Feb 2026 [23].

These climate planning instruments serve different but interconnected purposes. Their use in an integrated manner allows countries to leverage the synergies in planning and implementation over the short and long term. Countries can translate their long-term visions in LT-LEDS into actionable short- and medium-term goals in their NDCs. NDCs can be used to guide their NAP process, while NAPs provide the implementing strategies for turning NDCs commitments into on ground action. With proper planning and coordination, these instruments create a set of climate planning toolbox that provides a bridge between immediate actions to address climate change while being aligned with long-term sustainable development goals.

2.5.2 *Subnational Climate Action*

2.5.2.1 State Climate Initiatives

Climate action is not confined to national jurisdiction but can be supported or replaced by action from states and provinces. States such as California in the United States of America provide outstanding examples of continued leadership in climate action. California's cap-and-trade program, which is overseen by the California Air Resources Board (CARB) has set economy-wide targets to reduce the total amount of GHGs that can be emitted by major industries. The program has GHG reduction targets of 40% below 1990 levels by 2030, and goals to achieve carbon neutrality by 2045 [4]. Similarly in New York, the Climate Leadership and Community Protection Act (CLCPA) committed to 100% zero-emission electricity by 2040, sets legally binding standards to reduce GHG by 85%, and be completely carbon-neutral by 2050. It also mandates that 35% of the benefits of clean energy and energy efficiency must be routed to 'disadvantaged' communities, which are first and worst affected by the climate crisis [25].

The Regional Greenhouse Gas Initiative (RGGI), which is a regional CO_2 emissions reduction initiative from power plants in 11 states of the United States of America, is an example of how states can collaborate across boundaries. The participating states set a limit on CO_2 emissions, leading to co-benefits to their communities [28]. Over 15 states have adopted California's clean vehicles standards, demonstrating that subnational climate action can become a powerful catalyst for change despite changes in policies at the federal level [1]. These examples demonstrate that state initiatives can create a 'race to the top', as one jurisdiction sets ambitious targets which are matched by other states, accelerating climate action across entire regions.

2.5.2.2 City-Level Climate Initiatives

Apart from states, cities are also leading and contributing to climate action. C40 Cities, which is a global network of about 100 mayors of the world's leading cities, allows city administrators to share solutions and amplify their collective voice. Member cities of the network come together to raise climate ambition through supporting high-impact accelerators, fostering innovation, international advocacy and diplomacy, sharing best practices across high-impact sectors and facilitating projects to improve resilience in cities [2].

ICLEI—Local Governments for Sustainability (originally International Council for Local Environmental Initiatives) is a global network working with more than 2500 local and regional governments and is active in more than 125 countries. It works to create systemic change for urban sustainability through local action and helps cities and towns to become sustainable and climate resilient. It also promotes resource-efficient, biodiverse, low-carbon, equitable growth and a circular green urban economy. Members work together to promote low-emission development by driving local action through partnerships, capacity building, and advocacy [16].

The Global Covenant of Mayors for Climate and Energy is the world's largest global alliance for city and local governments' climate leadership, with over 13,500 cities as its members. The organization supports implementation of locally relevant solutions across mitigation, adaptation, and sustainable energy access through local initiatives, innovative financing models, and sustainable infrastructure [11].

2.5.2.3 Local Community-Based Approaches

Some of the impactful climate solutions emerge from local communities and grassroots initiatives, which create practical solutions to address environmental needs with broader climate goals. Community-based approaches rely on local resources and knowledge, making them more sustainable and acceptable than top-down interventions. Examples include community gardens that contribute to food security while sequestering carbon, community renewable energy cooperatives, or local waste management programs that reduce methane emissions. Participation of the community and local ownership give a sense of responsibility and control in designing and implementing solutions, which are more likely to succeed over time. The diversity of community-based approaches, which reflects local conditions and priorities, is a strong contributor to the success of the programs and complements the larger state-led climate initiatives.

Indigenous communities also bring rich experience in sustainable resource management and contribute to preserving nature, contributing actively to climate action. Traditional knowledge systems help in understanding long-term ecological sustainability, and nature-based solutions prove valuable to fill knowledge gaps while addressing social and cultural impacts. For example, Indigenous communities in the Arctic, rainforests of Latin America and nomads in Africa can help in climate observations to understand the long-term impacts of climate change, design traditional solutions to adapt to changes in climate, provide forest stewardship, contribute to community-led ecosystem monitoring and implement nature-positive approaches.

Indigenous-led organizations are increasingly playing a key role in climate governance, bringing perspectives that emphasize sustainability, nature-based solutions, long-term thinking, and the rights of future generations. The *UNESCO Climate Frontlines Initiative,* which is a global forum for Indigenous Peoples, small islands and vulnerable communities, recognizes this and works to integrate Indigenous and local knowledge into global climate science and policy processes [33]. The Local Communities and Indigenous Peoples Platform (LCIPP) of the UN Climate Change also brings together people and their knowledge towards achieving the objectives of the Paris Agreement [46]. UNESCO's Local and Indigenous Knowledge Systems (LINKS) promotes Indigenous and local knowledge in climate action by integrating it into policies and decision-making processes related to oceans, water, and land conservation through knowledge exchange workshops and advocacy [34].

2.5.3 Vertical and Horizontal Integration of Climate Governance

Vertical integration implies coordinating climate policies and plans across national, subnational and local governance to facilitate ambitious climate action. It involves creating linkages in planning, implementation, monitoring, evaluation and learning. Effective coordination mechanisms create pathways for information sharing, joint planning and collaborative implementation. Countries are experimenting with different approaches, such as forming intergovernmental committees and flexible working groups that bring together administrators from different levels. The effectiveness of vertical coordination also depends on the governance structures, autonomy and accountability of the bureaucracy. This requires clear roles and responsibilities for each level of government, streamlined mechanisms for the flow of information, collaboration, availability of adequate resources and manpower and monitoring systems.

Horizontal integration addresses the coordination across different government departments operating at the same governance level. This coordination is required among different ministries at the national level as well as the subnational level. Climate change cuts across sectors and hence coordination among energy, transport, industry, finance and environment ministries and departments is critical. The effectiveness of strategies is a function of policy alignment, and tradeoffs must be addressed. Without such alignment, these policies may undermine climate objectives due to conflicting goals.

The division of responsibilities as defined under the constitution is an important factor that impacts the achievement of horizontal climate policy integration and vertical coherence, especially in political governance systems which have autonomous departments and a high degree of independence between federal and subnational levels. For example, in Germany, the federal structure requires constant dialogue and discussion between national climate goals and state-level implementation capacity. In several other European countries, such as Austria and Switzerland, the central government exercises broad policymaking but responsibility for implementation rests with states. Brazil offers a different model, where all levels of government have policymaking powers for climate action. This allows independent action but also requires sophisticated coordination mechanisms. Brazil's climate governance approach includes partnerships among federal, state and local governments, the private sector, academia, and civil society, which demonstrate the comprehensive range of actions to operationalize vertical and horizontal integration in multi-level governance structures.

2.6 Global Climate Action Agenda and Non-state Actors

The need for involving all levels of society and a diverse set of actors was recognized early in the international climate governance process [47].

2.6.1 Lima-Paris Action Agenda

Following the convening of diverse climate stakeholders by UN Secretary General Ban Ki-moon in 2014, the Lima-Paris Action Agenda (LPAA) was established [35]. This was a joint initiative launched by France and Peru to mobilize governments and non-state actors to accelerate climate actions. The LPAA included governments, cities, regions, subnational entities, international organizations, civil society, Indigenous peoples, women, youth, academic institutions and businesses working individually and jointly to advance ambitious climate action [47]. At COP 21 in 2015, governments formally acknowledged that urgent climate action from all stakeholders was essential to achieve the goals of the Paris Agreement. The LPAA worked in parallel with official government negotiations to mobilize, engage and showcase the commitments of non-state actors to increase ambition, build momentum and stronger political will for climate action.

The Global Climate Action Agenda (GCAA) evolved from the LPAA and is a framework to facilitate climate action by non-state actors. It was launched at COP21 and complements national commitments under the Paris Agreement. Unlike Parties, non-state actors operate voluntarily, have higher ambition, move fast and adopt innovative approaches, making them an important contributor to climate action.

2.6.2 Marrakech Partnership for Global Climate Action

At the COP21, Parties also recognized the urgent need to bridge governmental efforts with the voluntary initiatives of non-state actors and formally agreed to position two Climate High-level Champions (CHLCs). They are appointed by the COP Presidency countries and hold the chair for two years. The CHLCs promote broader participation of non-state actors and link their efforts to countries' NDCs, thereby enhancing ambition and credibility to climate action.

The Marrakech Partnership for Global Climate Action (MPGCA) was established at COP22 under the leadership of the CHLCs, bringing together a diverse set of stakeholders across different sectors to enhance climate ambition [48]. The Marrakech Partnership addresses nine sectors—energy, industry, land use, oceans and coastal zones, water, human settlements, transport, resilience and finance—providing a platform for continued engagement of actors involved in implementation of climate action [48]. It does not substitute for negotiations among Parties but contributes to the implementation of the Paris Agreement by fostering greater ambition and enabling collaboration between national governments and non-Party stakeholders.

2.6.3 *Global Climate Action Portal*

The NAZCA (Non-State Actor Zone for Climate Action) platform is a global registry that tracks voluntary climate commitments by non-state actors such as businesses, cities, regions, investors, and civil society [49]. Launched in 2014, it was designed to provide a transparent database for reporting climate actions from both public and private sectors. By providing a clear, comprehensive view of global efforts, the platform highlights contributions, encourages accountability, and inspires greater climate ambition. It is renamed the *Global Climate Action Portal,* where over 49,000 actors have registered their climate commitments. By linking climate actions to broader climate goals, the portal boosts visibility, fosters collaboration, and encourages accountability, demonstrating how non-state actors can contribute to scaling global climate action.

2.6.4 *Business and Corporate Actions*

Businesses and corporates play an important role in global climate action beyond what governments mandate. Many of them have demonstrated setting ambitious, science-based targets, voluntary reporting of emissions, accounting for climate risks to enhance transparency, and influencing policy through corporate lobbying to support climate-friendly policies. This helps to create a global ecosystem where enhanced climate action is integral to business strategy.

2.6.4.1 Target Setting for Corporate Climate Commitments

The Science Based Targets initiative (SBTi) is an organization that drives corporate climate action by helping companies to set ambitious, science-based GHG reduction targets [29]. It provides guidance, tools, and independent validation needed to align with the Paris Agreement goal and to reach net-zero by 2050 at the latest. The SBTi framework provides a consistent and scientifically credible process for companies, financial institutions, and small and medium enterprises to advance their climate-related commitments. Through its Corporate Net-Zero Standard for different sectors, it holds companies accountable by requiring them to publicly report on their actions and provides independent target validation. This ensures that corporate climate commitments are credible and achievable in a time-bound manner. By end-2025, over 13,000 businesses across regions and industries had set emissions reduction targets, demonstrating the planned climate commitments of corporates and businesses.

2.6.4.2 Climate Reporting and Disclosure Frameworks

There has been a rapid mushrooming of Environmental, Social, and Governance (ESG) frameworks which have been adopted by corporates. These address a range of issues such as sustainability reporting, financial disclosure, climate risk disclosure, environmental disclosure and impact measurement frameworks, amongst others.

The Carbon Disclosure Project (CDP) is an independent environmental disclosure system [3]. It provides a standardized platform for cities, states, regions and other organizations to report their environmental data on emissions, energy, water, mitigation and adaptation actions through annual questionnaires, which is then used to score their performance. This improves transparency and accountability and enables stakeholders to make informed investment decisions. The Global Reporting Initiative (GRI) is another initiative which offers a comprehensive set of standards [13]. It provides a framework to assess and report ESG impacts. This allows transparent reporting and helps private and public organizations to communicate their performance in these areas. The International Sustainability Standards Board (ISSB) is an independent standard-setting body that develops sustainability-related financial disclosures. It was created by the IFRS Foundation to improve the consistency, comparability, and reliability of sustainability reporting [17]. As of date, ISSB has issued two standards that must be applied in conjunction for climate-related disclosures.

The IFRS S1—General Requirements for Disclosure of Sustainability-related Financial Information standard enables companies to disclose sustainability-related financial information by providing a framework to disclose material sustainability risks and opportunities across governance, strategy, risk management, and metrics/targets across the company's value chain. The IFRS S2—Climate-related Disclosures standard sets out specific requirements for disclosing climate-related risks and opportunities for companies. It mandates disclosures on physical and transition risks, uses scenario analysis to assess resilience covering the company's entire value chain. It follows the structure of the Task Force on Climate-related Financial Disclosures (TCFD), which was discontinued from 2023 [30].

There are other similar frameworks which support corporate climate action. European Sustainability Reporting Standards (ESRS) are applicable to companies covered by the EU's Corporate Sustainability Reporting Directive (CSRD). The Sustainable Finance Disclosure Regulation (SFDR) is applicable for financial market participants in the EU. The UN Global Compact (UNGC) also encourages companies to align with principles on human rights, labor, environment and anti-corruption.

Climate reporting and disclosure frameworks are often used in combination with target-setting frameworks and help drive global climate governance by standardizing how companies report their environmental impact. The frameworks also enable investors to make informed financial decisions based on the climate commitments and performance of companies.

2.7 Challenges and Governance Gaps

2.7.1 Coordination and Coherence Challenge

Non-binding targets and the lack of enforcement mechanisms of existing climate governance mechanisms are a major challenge. While the non-legally binding approach ensures near-universal participation, the reliance on voluntary target setting, such as voluntary defined NDCs and lack of penalties for not meeting the self-declared commitments continues to be a hurdle. Equally concerning is the delay in submission of voluntary NDCs by countries. This impacts the compilation of information and the synthesis report, and incomplete information hinders decision-making. This governance gap, due to the inbuilt reliance on diplomatic pressure rather than hard enforcement mechanisms, presents continuing challenges for the global climate governance mechanisms.

Climate governance involves multiple institutions and regimes that can have conflicting mandates and priorities. This overlapping climate governance across multiple sectors and institutions presents a coordination challenge. Some examples include the installation of wind turbines, which often have impacts on biodiversity conservation under the *Convention on Biological Diversity*. Climate policies such as subsidies for encouraging deployment of clean energy technologies are often against World Trade Organization rules, as they are discriminatory in nature. Trade barriers that apply additional tariffs on imports from countries with weaker climate policies, such as the EU Carbon Border Adjustment Mechanism (CBAM), are seen as protectionist measures by some developing countries. The presence of multiple actors also brings challenges of inadequate harmonization of climate actions, leading to gaps between pledges and actual climate action.

Competing priorities of different actors, political and economic division and uneven impact of the clean energy transition on different countries add to political economy constraints. Changes in political leadership sometimes lead to a complete reversal of country position wherein it moves from ambitious climate action to denial of climate change, risking long-term investments. Governments are also under pressure from the rising costs of climate action and its immediate impacts on taxpayers and have to balance it with the long-term benefits of investing in mitigation and adaptation measures. The short electoral terms coupled with the need for political survival often lead to actions which prioritize immediate economic relief over long-term climate action. The challenge is higher in a functional democracy where climate strategy hinges on changing public perception, voter margins and political stability. This prevents long-term multi-year climate commitments in democratic systems unless there is adequate public support for long-term climate action.

Trade-offs also exist between acting on climate change as compared to other Sustainable Development Goals (SDGs), leading to competing priorities. Examples include the impact of renewable energy production from large dams on river biodiversity, competition between use of land for food production or the generation of biofuels and the impact of coal mine closures on jobs. Lack of financial and other resources impacts the choice between mitigation and adaptation efforts, often creating a tradeoff for allocation of financial resources. Countries also have to balance competing priorities of economic growth, environmental protection and social development, which creates additional constraints for climate action.

2.7.2 *Equity and Justice Concerns*

The divide between the Global North and Global South is rooted in the historical responsibility of global GHG emissions. Equity, fairness and climate justice are demanded by the Global South in climate negotiations, while the Global North countries are concerned about the financial obligations and the ensuing liability claims. This leads to limited mobilization of resources and lack of coherence in negotiated outcomes.

One example is the operationalization of the *Fund for Responding to Loss and Damage* (FRLD) to address compensation for harm from climate change [10]. While the fund addresses some financial needs of vulnerable communities in developing countries to recover from loss and damage resulting from climate-induced extreme weather events, financial mobilization falls short of the needs arising from climate damages. Contributing nations also insist on calling the financial contributions voluntary 'solidarity' payments and not as 'reparations', which imply a form of legal obligation as compensation for climate harm. Nevertheless, concerns about climate justice continue to remain on account of the inadequacy of funding, liability of countries and administration of the fund.

The growing demand for a just transition for fossil fuel-dependent communities also reflects the implementation challenges for climate governance. A just transition implies fairness and inclusiveness for everyone concerned, creating decent work opportunities and leaving no one behind in the transition to a low-carbon and climate-resilient economy. This requires implementation of well-designed programs that address the impact of job losses for workers engaged in fossil fuel intensive industries such as coal mining. Measures such as reskilling of workers, providing income support and alternative job opportunities for affected regions must be prioritized to effectively address climate justice.

2.7.3 *Scale and Speed Mismatch*

There is a mismatch between the scale and speed of current action and what is required to meet the Paris Agreement goals. As seen from the outcomes of the first GST in 2023, despite overall progress, individual country commitments when aggregated fall short of collective action to meet the agreed targets [50]. The current NDCs put the world on track for a global temperature rise of 2.1–2.8 °C this century upon full implementation of NDCs [51]. However, the policies currently in place are insufficient to meet the existing NDCs [32]. Actions are required to cut global GHG emissions by 43% by 2030 and 60% by 2035 to get on track for net zero CO_2 emissions by 2050 and keep the 1.5 °C target within reach.

The call for action to achieve mitigation goals includes: tripling renewable energy and doubling energy efficiency; reducing non-CO_2 emissions including methane; transitioning away from fossil fuels in energy systems; and accelerating net zero emission energy systems. While some of these mitigation goals are being progressed, many of these are moving slowly. The window for action is fast closing, and there is a clear risk that the 1.5 °C target may soon slip out of reach.

There is a temporal disconnect between long-term climate cycles and short time horizons of democratic political systems. Climate change builds and unfolds over several decades and centuries, while power politics operates on much shorter election cycles of around 4–5 years. This creates political constraints on technologically feasible climate solutions. This often leads to symbolic implementation of softer policies, which have less impact, while strong measures that are costly and yield results in the long term are pushed forward or weakened considerably to avoid backlash. This mismatch between political incentives and physical atmospheric cycles contributes to the growing climate action gap.

On adaptation, it was recognized that considerable progress was made as countries submitted their NAPs. Gaps identified included the plans being fragmented, sector-specific, incremental, inadequately financed and unequally distributed across regions. Actions to bridge this gap included development of country adaptation actions, including establishing and operationalizing monitoring, evaluation and learning systems; implementation of integrated nature-based solutions and ecosystem-based approaches; and strengthening the linkages and synergies with the global goal on adaptation.

There is also a mismatch between the broad goals of the Paris Agreement, which is global in nature and actions on the ground which are more local and specific, creating a global—local divide. Cities, municipalities and elected authorities have local priorities such as job preservation, last-mile connectivity, cleanliness and adaptation, while countries and global organizations have larger and long-term goals, which creates misaligned agendas. Unequal resources, capacity and varying speeds of implementation lead to uneven progress in different communities, which further increases the gap between required actions and actual progress on the ground.

Scaling climate action requires a rollout of large-scale infrastructure such as electricity grids, energy storage technologies and clean energy infrastructure. Global energy systems and capital-intensive industrial infrastructure take several years to build and involve technology lock-in. Along with technologies, regulations, procedures and institutional processes are also required to support a functional system. Transforming centuries-old systems into a completely new system within decades is a continuing challenge which highlights the scale and speed mismatch between climate ambition and action.

2.8 Conclusion

Climate governance is evolving and requires a flexible approach. Centralized systems provide the framework but may lack the speed and granularity required to address climate change. Adaptive governance, which includes feedback, learning, and continuous improvement, is also required to make governance systems more acceptable and effective. There is a continuous demand for institutional reforms and innovations through strengthening the processes to advance international climate governance. The ETF enhances accountability for climate action and as these processes become embedded over time, they will lead to accurate measurement, clear reporting and focused support for advancing climate action.

Institutions must focus on implementation to deliver measurable results to bridge the gap between promises and climate action. These include financial institutions, intergovernmental development agencies, and non-state actors that support delivering finance, capacity building, technical assistance and developing public awareness to accelerate climate action. There is also a need for integration of multi-level governance in existing frameworks. Vertical integration includes linking local, subregional, national, regional, and global levels. Effective climate action also requires horizontal coordination across sectors and stakeholders, which breaks existing silos to advance coordinated planning. The role of Parties remains vital for setting NDCs, creating policies, and coordinating climate action across horizontal and vertical levels. Multiple decision centers with a well-defined role and effective coordination among actors are essential to tailor context-specific climate actions that are scalable. Building effective climate governance systems also requires participatory approaches, decentralized decision-making, and strong data-sharing mechanisms.

Building political momentum through strong coalitions and alignment of national climate agendas results in long-term continuity of climate policies. This can be strengthened by developing a social license for change through empowering communities. Building legitimacy for bottom-up climate action by identifying the beneficiaries and the specific benefits from climate action can help to mobilize long-term public and political support for climate governance. Integrating climate goals into regulations and making them legally binding makes climate action politically and economically irreversible. It also builds coalitions across geographies and political cycles, which reduces the risk of backtracking on committed actions. Climate governance frameworks that are adaptive, connect diverse actors and initiatives while allowing flexibility are most likely to succeed in the complex and evolving geopolitical environment.

References

1. Alternative Fuels Data Center (no date) Adoption of California's clean vehicle standards by state. https://afdc.energy.gov/laws/california-standards. Accessed 07 Mar 2025
2. C40 Cities (no date) About C40. https://www.c40.org/about-c40/. Accessed 10 Mar 2025

3. CDP (no date) CDP: turning transparency to action. https://www.cdp.net/en. Accessed 15 Mar 2025
4. California Environmental Protection Agency (2025) California's secretary for environmental protection joins state and global leaders at climate week NYC, urging climate action. https://calepa.ca.gov/2025/09/22/press-release-californias-secretary-for-environmental-protection-joins-state-and-global-leaders-at-climate-week-nyc-urging-climate-action/. Accessed 20 Mar 2025
5. Callendar GS (1938) The artificial production of carbon dioxide and its influence on temperature. Q J R Meteorol Soc 64(275):223–240. https://rmets.onlinelibrary.wiley.com/doi/abs/10.1002/qj.49706427503. Accessed 05 Mar 2025
6. Clean Air Fund (no date) The forgotten greenhouse gas: tropospheric ozone. https://www.cleanairfund.org/news-item/greenhouse-gas-tropospheric-ozone. Accessed 12 Mar 2025
7. Climate and Clean Air Coalition (no date) Short-lived climate pollutants. https://www.ccacoalition.org/content/short-lived-climate-pollutants. Accessed 12 Mar 2025
8. Climate and Clean Air Coalition (no date) Black carbon. https://www.ccacoalition.org/short-lived-climate-pollutants/black-carbon. Accessed 15 Mar 2025
9. Climate and Clean Air Coalition (no date) The climate and clean air coalition. https://www.ccacoalition.org/content/climate-and-clean-air-coalition. Accessed 10 Mar 2025
10. Fund for Responding to Loss and Damage (no date) Homepage. https://www.frld.org/. Accessed 17 Mar 2025
11. Global Covenant of Mayors for Climate & Energy (no date) Home. https://www.globalcovenantofmayors.org/. Accessed 13 Mar 2025
12. Global Methane Pledge (no date) Homepage. https://www.globalmethanepledge.org/. Accessed 19 Mar 2025
13. Global Reporting Initiative (no date) GRI: home. https://www.globalreporting.org/. Accessed 14 Mar 2025
14. International Civil Aviation Organization (no date) CORSIA eligible emissions units. https://www.icao.int/CORSIA/corsia-eligible-emissions-units. Accessed 11 Mar 2025
15. International Civil Aviation Organization (no date) Long-term global aspirational goal (LTAG) for international aviation. https://www.icao.int/environmental-protection/long-term-global-aspirational-goal-ltag-international-aviation. Accessed 12 Mar 2025
16. ICLEI–Local Governments for Sustainability (no date) Homepage. https://iclei.org/. Accessed 12 Mar 2025
17. IFRS Foundation (no date) International sustainability standards board. https://www.ifrs.org/groups/international-sustainability-standards-board/. Accessed 17 Mar 2025
18. International Maritime Organization (no date) Historic background. https://www.imo.org/en/ourwork/environment/pages/historic-background.aspx. Accessed 17 Mar 2025
19. International Maritime Organization (no date) The IMO net-zero framework: FAQs. https://www.imo.org/en/mediacentre/hottopics/pages/faqs-the-imo-net-zero-framework.aspx. Accessed 17 Mar 2025
20. International Air Transport Association (2026) Fact sheet: CORSIA. https://www.iata.org/en/iata-repository/pressroom/fact-sheets/fact-sheet-corsia/. Accessed 17 May 2026
21. International Energy Agency (no date) International shipping. https://www.iea.org/energy-system/transport/international-shipping. Accessed 17 Mar 2025
22. International Energy Agency (no date) Aviation. https://www.iea.org/energy-system/transport/aviation. Accessed 17 Mar 2025
23. NAP Central (no date) Submitted NAPs from developing country parties. https://napcentral.org/submitted-NAPs. Accessed 19 May 2026
24. National Institute of Water and Atmospheric Research (1985) Greenhouse gases and climate. UNEP/WMO/ICSU Tech Inf Circ 203. https://webstatic.niwa.co.nz/library/nzmstic203.pdf. Accessed 19 Mar 2025
25. NY Renews (no date) Climate leadership and community protection act. https://www.nyrenews.org/clcpa. Accessed 19 Mar 2025

26. Ozone Secretariat (no date) The Vienna convention for the protection of the ozone layer. https://ozone.unep.org/treaties/vienna-convention. Accessed 20 Mar 2025
27. Ozone Secretariat (no date) The montreal protocol on substances that deplete the ozone layer. https://ozone.unep.org/treaties/montreal-protocol/montreal-protocol-substances-deplete-ozone-layer. Accessed 20 Mar 2025
28. RGGI Inc. (no date) Intro to RGGI fact sheet. https://www.rggi.org/sites/default/files/Uploads/Fact%20Sheets/RGGI_Intro_Factsheet.pdf. Accessed 20 Mar 2025
29. Science Based Targets initiative (no date) Ambitious corporate climate action. https://sciencebasedtargets.org/. Accessed 21 Mar 2025
30. Task Force on Climate-related Financial Disclosures (no date) Homepage. https://www.fsb-tcfd.org/. Accessed 21 Mar 2025
31. United Nations Environment Programme (2023) Ozone layer recovery is on track, helping avoid global warming by 0.5 °C. https://www.unep.org/news-and-stories/press-release/ozone-layer-recovery-track-helping-avoid-global-warming-05degc. Accessed 21 Mar 2025
32. United Nations Environment Programme (2024) Emissions gap report 2024 press statement. https://www.unep.org/news-and-stories/statements/emissions-gap-report-2024-press-statement. Accessed 21 Mar 2025
33. United Nations Educational, Scientific and Cultural Organization (no date) Climate frontlines. https://www.unesco.org/en/links/climate-frontlines. Accessed 23 Mar 2025
34. United Nations Educational, Scientific and Cultural Organization (no date) Local and indigenous knowledge systems and climate change. https://www.unesco.org/en/climate-change/links. Accessed 23 Mar 2025
35. United Nations Framework Convention on Climate Change (2015) Paris Agreement. https://unfccc.int/sites/default/files/resource/parisagreement_publication.pdf. Accessed 23 Mar 2025
36. United Nations Framework Convention on Climate Change (no date) The compliance committee of the Kyoto Protocol. https://unfccc.int/Compliance-Committee-CC. Accessed 23 Mar 2025
37. United Nations Framework Convention on Climate Change (no date) The Paris Agreement. https://unfccc.int/process-and-meetings/the-paris-agreement. Accessed 23 Mar 2025
38. United Nations Framework Convention on Climate Change (no date) nationally determined contributions registry. https://unfccc.int/NDCREG. Accessed 25 Mar 2025
39. United Nations Framework Convention on Climate Change (no date) Conference of the parties (COP). https://unfccc.int/process/bodies/supreme-bodies/conference-of-the-parties-cop. Accessed 25 Mar 2025
40. United Nations Framework Convention on Climate Change (no date) Conference of the parties serving as the meeting of the parties to the Kyoto Protocol (CMP). https://unfccc.int/process/bodies/supreme-bodies/conference-of-the-parties-serving-as-the-meeting-of-the-parties-to-the-kyoto-protocol-cmp. Accessed 25 Mar 2025
41. United Nations Framework Convention on Climate Change (no date) Conference of the parties serving as the meeting of the parties to the Paris Agreement (CMA). https://unfccc.int/process/bodies/supreme-bodies/conference-of-the-parties-serving-as-the-meeting-of-the-parties-to-the-paris-agreement-cma. Accessed 25 Mar 2025
42. United Nations Framework Convention on Climate Change (no date) Subsidiary body for scientific and technological advice (SBSTA). https://unfccc.int/process/bodies/subsidiary-bodies/sbsta. Accessed 25 Mar 2025
43. United Nations Framework Convention on Climate Change (no date) Subsidiary body for implementation (SBI). https://unfccc.int/process/bodies/subsidiary-bodies/sbi. Accessed 26 Mar 2025
44. United Nations Framework Convention on Climate Change (no date) Long-term strategies portal. https://unfccc.int/process/the-paris-agreement/long-term-strategies. Accessed 26 Mar 2025
45. United Nations Framework Convention on Climate Change (no date) National adaptation plans. https://unfccc.int/national-adaptation-plans. Accessed 26 Mar 2025

46. United Nations Framework Convention on Climate Change (no date) Facilitative working group of the local communities and indigenous peoples platform (FWG). https://www4.unfccc.int/sites/NWPStaging/Pages/Facilitative-Working-Group-of-the-Local-Communities-and-Indigenous-Peoples-Platform.aspx. Accessed 26 Mar 2025
47. United Nations Framework Convention on Climate Change (no date) Global climate action. https://unfccc.int/climate-action. Accessed 26 Mar 2025
48. United Nations Framework Convention on Climate Change (no date) About the Marrakech partnership. https://unfccc.int/climate-action/events/about-the-marrakech-partnership. Accessed 27 Mar 2025
49. United Nations Framework Convention on Climate Change (no date) Global climate action portal. https://unfccc.int/climate-action/tracking-and-recognition/global-climate-action-portal. Accessed 27 Mar 2026
50. United Nations Framework Convention on Climate Change (no date) Outcome of the first global Stocktake. https://unfccc.int/topics/global-stocktake/about-the-global-stocktake/outcome-of-the-first-global-stocktake. Accessed 27 Mar 2026
51. United Nations Framework Convention on Climate Change (no date) GST outcomes: mitigation (image). https://unfccc.int/sites/default/files/2024-09/GST_Outcomes_MITIGATION.png. Accessed 28 Mar 2026
52. United Nations (1997) Kyoto Protocol to the United Nations framework convention on climate change. https://unfccc.int/resource/docs/convkp/kpeng.pdf. Accessed 28 Mar 2025
53. United Nations Development Programme (2022) Volume III: what is climate governance? Toolkit for young climate activists. https://www.undp.org/sites/g/files/zskgke326/files/2022-09/Volume%20III%20What%20is%20climate%20governance.pdf. Accessed 03 Mar 2025
54. World Meteorological Organization (no date) World climate programme (WCP). https://community.wmo.int/en/world-climate-programme-wcp. Accessed 03 Mar 2025
55. World Meteorological Organization (1989) The changing atmosphere: implications for global security, Toronto, Canada, 27–30 June 1988–conference proceedings. Secretariat of the World Meteorological Organization, Geneva. https://cmosarchives.ca/History/ChangingAtmosphere1988e.pdf. Accessed 03 Mar 2025
56. World Meteorological Organization and United Nations Environment Programme (1992) Climate change: the IPCC 1990 and 1992 assessments. WMO/UNEP, Geneva. https://www.ipcc.ch/site/assets/uploads/2018/05/ipcc_90_92_assessments_far_full_report.pdf. Accessed 28 Mar 2025

Chapter 3
Climate Change Policies: A Tool to Deliver Positive Climate Outcomes

Abstract This chapter positions climate change policies as a tool to deliver positive climate outcomes. Different types of climate policy instruments are classified and presented using examples from different sectors and countries. This includes regulatory approaches, economic instruments, voluntary and cooperative approaches and information, education and capacity development policies. The chapter emphasizes the need for adopting complementary instrument mixes and provides practical examples of implementing successful policy packages. It also discusses the barriers to policy implementation and presents frameworks and methods for measuring policy effectiveness.

Keywords Climate policy · Regulatory approaches · Economic instruments · Carbon pricing · Instrument mix · Policy effectiveness

3.1 Introduction

Global climate governance frameworks need to be operationalized into on-the-ground climate action and measurable outcomes through well-designed and effective climate policies. Climate policies provide strategic direction and a long-term signal of the intent of the government, which can serve as a powerful tool to drive climate action. Well-designed climate policies can correct market failures, mobilize climate finance, accelerate technology development for climate gains, build climate momentum and deliver co-benefits beyond climate. Climate policies therefore act as a bridge between climate ambition and climate-positive outcomes and must be used effectively by governments to drive climate action.

This chapter highlights the important role of climate change policies as a tool to deliver positive climate outcomes. It commences with classifying climate policy instruments and defining the multi-actor climate policy landscape. Section 3.2 presents the regulatory approaches, including different forms of directives and standards, using global examples. Section 3.3 describes economic instruments such as financial support measures, fiscal incentives, carbon pricing and market-based

K. Narula, *The Climate Catalyst*, Advances in Sustainability Science and Technology,
https://doi.org/10.1007/978-981-92-5213-8_3

approaches using examples from different sectors. Section 3.4 describes voluntary and cooperative approaches and Sect. 3.5 presents information, education and capacity development policies to advance climate action. Section 3.6 emphasizes the complementarity of instrument mixes and practical aspects of implementing policy packages using different examples. Section 3.7 reflects on the barriers to policy implementation. Measuring policy effectiveness is critical for testing policies, and several frameworks and methods are discussed in Sect. 3.8, followed by the concluding section.

3.1.1 Policy as a Catalyst for Transformation

Climate policy refers to the overarching framework of interventions designed to manage and respond to climate change. A well-designed climate policy addresses both mitigation of GHG emissions and adaptation to the impacts of climate change. Climate policy can become a driver of transformation as it can reshape economic systems and trigger a systemic shift. Policies provide a tool to deliver the desired outcomes as decision makers can define rules and set guidelines. Well-designed policies guide the respondents to choose those options that lead to the desired outcomes, among a set of alternatives. Policies can deliver specific goals as rules, and the framework can be designed to address different barriers. Effectively implemented policies can result in minimal friction and can lead to a systemic transformation.

3.1.2 Classification of Policy Instruments

There are several ways to classify climate policies and they essentially work on the principles or 'carrot' or 'stick'. The 'carrot' approach is a positive incentive or a form of reward which motivates actors to undertake a desired action, creating a 'pull' factor. In contrast, a 'stick' approach creates the fear of a negative outcome which drives actors to avoid the negative consequence of their actions, thereby providing a 'push' factor. Often, both these principles are used in tandem as they address different actors with different behaviors.

There are several taxonomies for classification of climate policy instruments. The OECD broadly classifies them into market-based instruments and non-market-based instruments [10]. Market-based instruments are designed to influence actors' behavior by offering choices that have different economic incentives. On the other hand, non-market-based instruments work through non-monetary incentives to change behavior, such as through mandatory regulations and obligations [15]. Other classification frameworks prefer to use the nomenclature of regulatory and economic instruments for climate policies [6]. Voluntary and information-based instruments are acknowledged as being distinct from the above-mentioned approaches and are sometimes clubbed together in climate policy taxonomies [11].

3.1.3 The Multi-actor Policy Landscape

Climate action involves multiple actors, which include governments at different levels (municipality, city, state, federal), the private sector, households and individuals. Different actors have different motivations and limitations, and hence there is no one-size-fits-all policy. Different climate policy instruments address different challenges and have to be tailored to overcome specific hurdles. To be effective, climate policies require a coordinated approach across different levels, sectors, and actors while ensuring that policies are mutually reinforcing. This multi-actor policy landscape makes policy design complex and context-dependent.

3.2 Regulatory Approaches

Regulatory approaches are also called the command-and-control approaches and are a form of non-market-based instruments. In the regulatory approach, the rule setter introduces directives and standards that have a legal backing and may have penalties for non-compliance.

3.2.1 Directives

3.2.1.1 Mandates

Table 3.1 provides a description and examples of different policy instruments where mandates are used by the government to drive climate action.

3.2.1.2 Transparency-Based Directives

These directives aim to improve transparency by providing information about the product to consumers. It relies on making facts accessible to stakeholders, for example, through mandatory requirements to make environmental performance data public. This creates external pressure from investors and consumers of the company to improve their environmental performance. Table 3.2 provides a description and examples of different policy instruments where transparency-based directives are used to drive climate action.

Table 3.1 Policy instruments and examples of mandates

Instrument	Description	Examples
Ban on technology and sale of specific product	Prohibit the sale or installation of high-emission technologies and products	Gas boiler bans in Austria and the Netherlands for building heating Sale of incandescent bulbs banned Coal power plant phase-out in Germany and Canada Ban on use of plastic bags in Kenya for environmental protection
Sunset provision	Scheduled phase-out of high-emission technologies giving time for a planned transition	Scheduled coal phase-out in Canada and France HFC phase-down to 85% by 2036 for developed countries Progressive phase-out of gas-based boilers in Greece and Denmark Internal combustion engine vehicle sales phase-out in the EU, UK and Canada
Renewable energy obligations	Mandates a minimum share of renewable energy in the overall energy mix	Renewable portfolio standards (RPS) for minimum 40% share of renewable energy by 2030 in states of the U.S Solar photovoltaic installation mandated for new low-rise residential buildings in California Renewable electricity procurement mandate in China and India for industries 15% biodiesel blending mandate in Brazil for fuels

3.2.1.3 Other Regulations

Table 3.3 provides a description and examples of policy instruments using different forms of regulations.

3.2.2 Standards

While regulations are legally enforceable, standards are voluntary guidelines or specifications that describe best practices. Standards can also become mandatory if they are incorporated as a regulation.

Table 3.2 Policy instruments and examples of transparency-based directives

Instrument	Description	Examples
Mandatory reporting and disclosure	Companies are required to measure and report emissions or environmental performance in standardized data formats	Climate risk disclosure is mandatory in the UK, New Zealand and the UAE for large companies Annual GHG reporting program by large emitters in the US Mandatory audited disclosures on ESG under corporate sustainability reporting directive (CSRD) in the EU for corporates Mandatory energy audits for large companies under EU energy efficiency directive
Mandatory labeling	Providing information on the product's environmental attributes to influence consumer choice	Energy performance certificates (EPCs) mandatory for selling and renting buildings in the EU Vehicle fuel economy labels are mandatory in new light-duty passenger vehicles in the EU, China and the US Product carbon footprint label is mandatory for rechargeable industrial batteries over 2 kWh in the EU
Monitoring, reporting and verification	Systematic process of measuring environmental parameters, formally submitting data to an authority, and having the data validated by an independent third party	Continuous emissions monitoring systems are mandatory in power plants in China, the US and India Third-party audits for emission data and sustainability reporting are mandatory in the EU UAE introduced emission reporting requirements and submission of verified disclosures for high-emitting companies Soil carbon monitoring requirements under implementation in the EU

3.2.2.1 Performance and Emissions Standards

These standards work through setting the limit on the amount of GHGs that can be emitted per unit of output, activity, or product. Complying with these efficiency or emission standards requires the use of specific technologies or practices.

Table 3.3 Policy instruments using different forms of regulations

Instrument	Description	Examples
Extended producer responsibility (EPR) rules	Producers have to take responsibility for the complete life cycle of a product, including recycling and final disposal after use, which creates a financial incentive for making products more eco-friendly, durable and recyclable	Mandatory recycling or take-back schemes for batteries, electrical and electronic equipment are applicable in the EU Recyclable packaging obligation with a minimum recycled content in the EU
Permitting and licensing	Set legally enforceable conditions to operate a facility based to comply with emission requirements	Mandatory environmental impact assessments (EIA) for large infrastructure projects to prevent long-term environmental damage Operating permits that cap emissions per facility under the US Clean Air Act Approval conditions with strict thresholds for emission of air pollutants for new industries
Land use and zoning regulations	Laws to designate and restrict the permissible uses of land which have an indirect impact on emissions	Land zoning encourages compact urban planning and limits the need for long transits Mandates for green spaces and restrictions on urban deforestation or land conversion build climate resilience

Table 3.4 provides a description and examples of different policy instruments where performance and emission standards are used.

3.2.2.2 Process and Practice Standards

Table 3.5 provides a description and examples of different policy instruments for process and practice standards to drive climate action.

3.2.2.3 Market Access and Procurement Standards

Table 3.6 provides a description and examples of different policy instruments for providing access to markets and for procurement.

Table 3.4 Policy instruments and examples of performance and emission standards

Instrument	Description	Examples
Performance standards	Sets the minimum level of target to be achieved without prescribing technology. This allows flexibility and promotes innovation for cost-effective compliance	Emission from all new power plants limited to 355 kg CO_2e/MWh on an annual basis as per Tier 1 standard in Singapore
Product standards	Regulates the characteristics of products and appliances sold by setting requirements for the environmental attributes of a manufactured product	Corporate Average Fuel Economy (CAFÉ) standards expressed in miles per gallon for different vehicles in the US Minimum standard of 120 lumens per watt for general service lamps in the US from 2028
Energy efficiency standards	Sets minimum energy efficiency level to lower energy demand	Building energy codes like ASHRAE standard for commercial buildings specifying different energy efficiency values such as for insulation Progressively improving electric motor efficiency standards from international energy efficiency class IE1 to IE5 in the EU Minimum energy efficiency ratings for appliances like refrigerators and air conditioners
Technology standards	Sets strict directives for the use of specific technologies or equipment	All coal power plants must install scrubbers to reduce SOx emissions Use of catalytic converters in vehicles to reduce emissions of unburnt hydrocarbons and black carbon
Cap on total emissions	Fixed limit on total emissions from a specific source or a sector over a specified period	The overall EU-wide cap for GHG emissions from fixed installations was fixed at 2,084 $MtCO_{2e}$ (2013) reducing to 1,816 $MtCO_{2e}$ (2020) under the EU ETS
Emission intensity standards	A form of performance standard on a facility's effluent which limits emission intensity but not total emissions	Starting 2023, coal plants are allowed to emit 260 mg/cu. mts. of SO_2 in Thailand The average emissions of new passenger cars sold by manufacturers in the EU should be lower than 93.6 gms CO_2 per km (2025–29)

(continued)

Table 3.4 (continued)

Instrument	Description	Examples
Ambient quality standards	Sets a minimum acceptable level of environmental quality, such as for air, to monitor overall impact without controlling specific sources	PM2.5 and NO_2 concentration limits in urban cities National air quality standards
Fuel standards	Sets limits for the physical or chemical composition of fuels, e.g., maximum sulfur content	The maximum sulfur content allowed in Canada for road transport diesel vehicles is 15 ppm (mg/kg)

Table 3.5 Policy instruments and examples of process and practice standards

Instrument	Description	Examples
Best available technology (BAT)	Require use of best proven technologies in each industry	EU industrial emissions directive requires industries to use BAT for pollution control. BAT reference values are documented for various industrial activities
Management system standards	Voluntary environmental and energy management processes such as ISO 14001 and ISO 50001 are mandatory for a specific set of companies in certain jurisdictions	Under the Energy Efficiency Act (EnEfG), Germany requires companies having total annual final energy consumption above 7.5 GWh to implement an energy management system in accordance with ISO 50001 or an environmental management system from July 2025
Operational standards	Regulates the process of conduct of different activities	Tillage and fertilizer application limits in agriculture sector. Under the EU Nitrates Directive, there is a cap on application of 170 kg of Nitrogen per hectare per year in designated 'Nitrate Vulnerable Zones' (NVZs) Ban on destructive methods such as use of explosives to catch fish, ban on bottom trawling and limits on use of minimum mesh sizes and length of fishing nets

3.3 Economic Instruments

Economic policy instruments create financial motivation for stakeholders and act as a 'pull factor' which encourages desired behavior from actors. They comprise financial support mechanisms, fiscal incentives and carbon pricing mechanisms.

Table 3.6 Policy instruments and examples of market access and procurement standards

Instrument	Description	Examples
Public procurement standards	Minimum environmental requirements for purchase of goods by governments	Zero-emission procurement policy for deployment of 10,000 electric buses via a Public–Private Partnership (PPP) model under PM-eBus Sewa in India Under the Buy Clean California Act, only low-emission steel can be procured for public works
Border adjustments	Additional tariffs on goods crossing borders based on carbon content	EU Carbon Border Adjustment Mechanism (CBAM) requires importers to pay the differential carbon price on specific goods
Market entry requirements	Use of minimum standards for accessing markets in a specific jurisdiction	Technical screening criteria such as minimum requirements for building energy efficiency is applied before providing market access for real estate investors Kenya has banned the import of used vehicles older than 8 years

3.3.1 Financial Support

3.3.1.1 Subsidies and Grants

The cost of new technologies is often higher than that of the existing alternatives. Subsidies and grants bridge this gap partially or fully by lowering the cost of low-carbon technologies for the consumer (Table 3.7).

3.3.1.2 Loans

These financial support mechanisms are designed to lower financing barriers by reducing borrowing costs and de-risking investments to attract private investment, thereby improving project bankability. Table 3.8 provides a description and examples of different policy instruments promoting loans to drive climate action.

3.3.1.3 Research, Demonstration and Deployment Grants

These targeted measures accelerate innovation and commercialization of low-carbon technologies. Public funding provided through these grants reduces financial and technical barriers, de-risks innovation, strengthens national innovation ecosystems and supports deployment of clean technologies. Table 3.9 provides a description and examples of different policy instruments for research grants.

Table 3.7 Policy instruments and examples of subsidies and grants

Instrument	Description	Examples
Feed-in tariffs (FITs)	FITs guarantee renewable energy developers an additional price for electricity that is fed into the grid, which ensures guaranteed returns for the developer, enabling investment	In 2000, the German government started a FIT for 20 years for each kWh of renewable electricity injected into the grid Wind projects in Denmark received a fixed price FIT for 20 years Kenya provided FITs for small hydropower for 15 years The UK provided FIT for anaerobic digestion and micro combined heat and power (up to 2 kW) for 10–25 years
Capital subsidies	Initial capital costs for installing clean technology are provided as a fixed percentage of total cost or a specified amount to overcome the investment cost barrier	India provides capital subsidy for residential rooftop solar plants under the 'PM Surya Ghar-Muft Bijli Yojana' Netherlands provides capital subsidies for business owners to install grid-connected wind turbines, which covers around 20% of the total investment France provides an ecological bonus (about 27% of the cost, up to a maximum of €4,000) for purchasing new electric passenger cars based on individuals' income levels
Investment grants	These grants are often one-time and provide direct funding for the acquisition of assets for covering part of a project cost to attract private co-investment	The EU Innovation Fund offered investment grants to pilot or small-scale industrial projects for green hydrogen and Carbon Capture and Storage (CCS) projects Grants were provided to fund climate-smart irrigation systems in Kenya and Malawi The U.S. Department of Energy (DOE) provided grants for deployment of large-scale public Electric Vehicle (EV) charging infrastructure for medium-and heavy-duty EVs

3.3.2 *Fiscal Incentives*

Fiscal incentives use the taxation system to encourage behavior change toward adoption of low-carbon and climate-resilient investment. They create an incentive for the company or individual users by reducing taxes, duties and other forms of levies

Table 3.8 Policy instruments and examples of loans

Instrument	Description	Examples
Preferential loans	These loans are offered for specified technologies at below-market interest rates or with longer repayment periods	Germany's KfW (German development bank) renewable energy program provides low interest rate (1–2%) loans for renewable energy projects China offers interest rate discounts for undertaking energy-efficient building renovations for developers and property owners
Soft loans	These are concessional loans with extended repayment terms given by international donors or multilateral banks to governments for specific projects	International Fund for Agricultural Development (IFAD) offers soft loans to small landholders for climate-resilient irrigation and crop diversification KfW provides soft loans for infrastructure development in emerging economies
Loan guarantees	Government agencies back private loans as a guarantor, which reduces lenders' risk, enabling access to capital	The World Bank's scaling solar program provides loan guarantees for African solar projects through the Multilateral Investment Guarantee Agency (MIGA) The U.S. DOE Loan Programs Office provided loan guarantees for CCS, electric vehicles and battery manufacturing projects

on clean technologies or by imposing them on polluting technologies. Table 3.10 provides a description and examples of different types of fiscal incentives.

3.3.3 Carbon Pricing

Carbon pricing is an economic policy instrument that uses a price on carbon to encourage a reduction in emissions. It works by creating incentives for companies and individuals to lower emissions by putting an additional cost on the use of fuels with higher carbon content. This financial burden acts as a motivator for companies to invest in cleaner technologies, energy efficiency and other control measures to limit emissions. At the end of 2025, over 52 countries and 43 subnational governments had implemented or were considering carbon pricing instruments.

Carbon emissions have significant externalities such as climate change and this cost is borne by the public. Carbon pricing mechanisms internalize the cost of carbon emissions and make emitters pay for them. When implemented successfully, it results in aggregate lower costs of controlling emissions for society. In many cases, the revenue from carbon pricing is used to fund public investments and to reduce health-care expenses for citizens. Carbon pricing can be divided into tax-based approaches

Table 3.9 Policy instruments and examples of research grants

Instrument	Description	Examples
Research and development (R&D) grants	Early-stage research is funded to reduce technical risks	The horizon europe programme provides funds for R&D and innovative projects to address climate change through competitive grants The U.S. DOE ARPA-E program funds early-stage research for next-generation battery technologies Japan's New Energy and Industrial Technology Development Organization (NEDO) supports private R&D projects on renewable energy integration and smart grids FAO funds R&D in climate-smart crop varieties and precision irrigation in the agriculture sector
Technology demonstration and deployment support grants	These grants finance pilot and large-scale demonstration projects to prove commercial viability of new technologies to bridge the gap between research and commercial deployment	The EU Innovation Fund provides grants for first-of-a-kind projects in CCS and steel decarbonization Germany's H2Global program co-funds green hydrogen pilot plants India's National Solar Mission financed pilot projects for concentrated solar power (CSP) projects Canada provided partial funding for CCS demonstration at industrial sites

and market-based approaches and its choice depends on national circumstances and specific policy objectives.

3.3.3.1 Tax-Based Approaches

Carbon taxes and CBAM are tax-based approaches to carbon pricing compliance mechanisms.

Table 3.10 Policy instruments and examples of fiscal incentives

Instrument	Description	Examples
Tax credits	Lower tax liability for investment in clean technologies	U.S. Residential Clean Energy Credit gives a 30% tax credit of the costs of new, qualified clean energy equipment such as solar water heaters, fuel cells and battery storage Reduced property tax rates for LEED certified homes in the U.S
Tax deductions and value added tax (VAT) exemptions	Lower VAT, customs duties, excise taxes on clean technologies and fuels	Brazil has a zero tax on imported solar modules and inverters Norway exempts electric vehicles from registration tax and has lower VAT Biofuels receive tax exemptions in Sweden
Environmental taxes	They impose charges on activities that harm the environment	EU applies tax on non-recycled plastic waste Japan's green tax penalizes high-emission cars that emit beyond the emission standards
Accelerated depreciation	Allows businesses to reduce their taxable income upfront and faster write-offs for clean technologies, improving the payback period	India and South Africa permit 40–50% accelerated depreciation for the first year on solar and wind assets

Carbon Taxes

Carbon taxes are imposed on designated entities, which are required to pay a tax according to their emissions. An explicit tax rate is set by the government based on the carbon content of fossil fuels in terms of price per tCO_{2e}. This fixed price imposed on emissions is often applied to the power sector and high-emitting industry while leaving out smaller sources of direct and indirect emissions such as vehicles. It is relatively simpler to administer as the emissions can be quantified and verified based on the fuel consumption. The tax is collected by the government on a yearly basis and can be used for climate-related activities.

For example, in Canada, the federal fuel charge sets specific rates for different types of coal used in electricity production in provinces such as Ontario, Manitoba, Saskatchewan and Alberta. For the period from April 1, 2024 to March 31, 2025, the imposed tax rates for high heat value coal were C$178.48 per tonne, while for low heat value coal it was C$141.80 per tonne. The specific rate is calculated based on the carbon content and standard emission factors for coal, and the above rate reflects a carbon price of C$80 per tonne of carbon dioxide equivalent (CO_{2e}) in 2024 [1].

As the tax is imposed at a fixed rate without any cap on emissions, it provides an additional revenue stream for the government, but total emissions may continue to grow as there is no cap on emissions. For example, in the case of a carbon tax on coal, the total amount of emissions depends on the electricity generated using coal and emissions could increase over time, despite the tax. Hence, the emission reduction outcome of a carbon tax is uncertain, but the carbon price is fixed for a particular duration. The Canadian government has announced an increase in the fuel charge by C$15 per tonne annually from 2024 to reach C$170 per tonne in 2030, doubling the carbon tax in six years. This provides certainty for making investment decisions in clean technologies in the long term.

As carbon tax is simple to implement and raise tax revenues, they have been implemented in 43 jurisdictions and cover 5% of global GHG emissions by volume. A carbon tax can be applied to different sectors within the country and also at the subnational level. At approximately $132 per tonne of carbon dioxide equivalent (CO_{2e}) in 2024, Switzerland had one of the highest carbon taxes in the world. The range varies widely across countries, with Poland imposing as low as $0.10 per tonne of CO_{2e} in 2024 [20]. Globally, carbon taxes contributed to about $33 billion in annual revenue for governments.

Carbon Border Adjustment Mechanism

A Carbon Border Adjustment Mechanism (CBAM) is a tariff levied on imported goods based on the GHG emissions embedded in their production. By imposing an additional price equivalent to the domestic carbon price (e.g., the EU Emissions Trading System allowance price) on embodied carbon in the imported goods, the CBAM aims to provide a level playing field to domestic producers. The CBAM functions as a form of carbon pricing mechanism, as the importer of goods must purchase additional 'CBAM certificates'. This eliminates the cost differential and equalizes the price of carbon for producing the goods under a carbon pricing mechanism and those produced outside the jurisdiction where the carbon pricing mechanism is enforced. CBAM also helps to address 'carbon leakage', a phenomenon where businesses shift industrial production from jurisdictions with high carbon prices to countries with lower or no carbon pricing. This relocation of production simply shifts the geographic location of emissions to avoid carbon costs, without reducing the overall global emissions. The CBAM attempts to mitigate the risk of carbon leakage and also encourages trading partners to adopt some form of carbon pricing mechanisms.

European Union has implemented the CBAM from 01 January 2026, which covers several sectors including iron & steel, cement, aluminum, fertilizers, electricity and hydrogen. From 2023 to 2025, the EU CBAM was in its transitional phase [4]. The UK Carbon Border Adjustment Mechanism (CBAM) is planned to be implemented from 01 January 2027 for all the above sectors except electricity [5].

3.3.3.2 Market-Based Approaches

Emissions Trading Systems

An Emission Trading System (ETS) is a 'cap and trade system' where a limit is set on total emissions by a sector or an industry. Companies and businesses are then allowed to buy and sell emission permits, which creates a market-determined price for carbon. Companies that can reduce emissions below their allocated limit are allowed to sell their extra allowances to companies that exceed their allowance limit. This provides certainty on the total volume of emissions, as aggregate emissions are restricted within the set cap for the industry, making it a useful tool to achieve climate commitments for a country.

As the price of carbon is a function of supply and demand, it is dynamic and could vary over the compliance cycle. This variable price may create budget uncertainty for businesses that rely on buying these credits over the long term. Price safeguards such as a price floor, price ceiling and market stability reserve are often implemented to limit high price volatility. Well-functioning markets also require a sufficient number of market participants for liquidity and hence it is difficult to implement this mechanism in countries with a limited number of buyers and sellers. To overcome this limitation, carbon offsets and linking of ETS with other jurisdictions are often exercised to increase tradability and market liquidity. Linking of different ETS also allows additional benefits from lower abatement costs for buyers and investment flow for sellers. Overall, this mechanism ensures that GHG emissions cuts happen where they are most cost-effective.

ETS is relatively more complex to implement due to higher administrative costs of monitoring, reporting, and verification (MRV) and technical complexities in setting the allowances and overall cap. Many jurisdictions start with a pilot ETS or a voluntary phase and then implement the full ETS in a phased manner. ETS is not suited for small-scale dispersed sources of GHG emissions such as cars and vehicles and is often restricted to large point sources of emissions such as industry and electricity producers.

At the end of 2025, around 37 ETS had been implemented globally, which covered about 23% of global GHG emissions by volume. At approximately $74 per tonne of carbon dioxide equivalent (CO_{2e}) in 2025, the EU ETS had the highest average price in the world. On the other end of the spectrum, Indonesia ETS had the lowest price of $1 per tonne of CO_{2e}, showing the large variation in prices [18]. The revenue earned globally in 2025 from ETS was approximately $80 billion, which was about three times that earned from carbon taxes [19].

Carbon Crediting Mechanisms and Carbon Offsets

Carbon crediting mechanisms are market-based approaches to reduce GHG emissions. A carbon credit represents verified emission reduction, avoidance, or removal of one tonne of CO_2 or its equivalent GHG emissions. Independent third parties like Verra, which manages the Verified Carbon Standard (VCS) program, Gold Standard, American Carbon Registry (ACR), and Climate Action Reserve (CAR) provide frameworks to certify and issue carbon credits for projects to reduce or remove carbon. After emission reductions are verified against specific standards' criteria, credits are issued, which can then be traded, or companies can use them to offset their own emissions. After a credit is used to offset emissions, it is 'retired' or permanently removed from circulation so that it cannot be reused.

Carbon crediting mechanisms include defining a baseline level of emissions, and only projects that reduce emissions below the baseline are eligible for the issuance of carbon credits. The projects also require robust MRV processes following approved methodologies. Further, it is important to establish the 'additionality', i.e., it must be established that the reduction or removal would not have occurred without the financial incentive from the sale of carbon credits. The issued carbon credits can feed into multiple systems such as voluntary markets or compliance markets, domestically or internationally. Carbon crediting mechanisms can enable climate finance for carbon mitigation projects such as renewable energy generation, reforestation, methane capture and improvements in energy efficiency, especially in emerging economies, as a steady revenue stream can be earned from selling carbon credits.

Carbon offsets are carbon credits that a company or an individual can buy to compensate for their own emissions. They help in meeting voluntary climate targets or compliance obligations for companies and enhance their sustainability and Environmental, Social, and Governance (ESG) credentials. The price of carbon credits in voluntary markets is influenced by several factors, including the project type, location, quality of credits and co-benefits beyond emission reduction. Several companies currently use carbon offsets to compensate their corporate emissions. Carbon Offsetting and Reduction Scheme for International Aviation (CORSIA) is a well-known global offsetting scheme to lower emissions from international aviation.

3.3.3.3 Hybrid Systems

Countries and subnational governments often implement a combination of carbon tax and an ETS to suit the national circumstances. Carbon taxes and ETS co-exist in 24 jurisdictions. Both instruments together covered about 15 $GtCO_2e$, which was about 28% of global GHG emissions. Often a country first adopts a carbon tax on fossil fuels, followed by a pilot ETS. In a hybrid system, both carbon taxes and a full-fledged ETS exist simultaneously.

For example, if carbon pricing instruments are applied in the power sector in a country, the ETS could be implemented for power stations that emit above a certain pre-decided threshold and a carbon tax could be implemented for smaller facilities that are lower than this threshold. This threshold is 26,000 tCO_{2e} in China, which covers nearly all coal plants over 50 MW generation capacity. Sweden follows a hybrid system where a carbon tax covers transport, heating, and residential use. In parallel, the EU ETS is applicable for large industrial and power sector entities. As the system is complementary, there is no overlap between sectors to avoid double taxation on carbon. Often the hybrid system is designed such that gradually the coverage of ETS increases to other sectors.

3.3.3.4 Internal Carbon Pricing

Internal Carbon Pricing (ICP) is used by individual companies to assign a monetary value to their own carbon emissions. Companies use this internal price for strategic decision-making, such as evaluating the viability of new projects based on future carbon prices to guide investment and operational decisions. ICP allows for assessing the risks related to future regulations, including regulatory changes such as the implementation of carbon taxes and a compliance ETS in a specific jurisdiction. This shadow price of carbon is a specific form of ICP where the hypothetical cost of carbon is included in planning and budgeting. The shadow price is a notional value which is incorporated into financial modeling and analysis and no money is exchanged.

An internal carbon fee involves a company charging its internal business units a fee based on their emissions, which creates a financial incentive for units to reduce their GHG emissions. The revenue generated from these internal fees is often pooled and reinvested in GHG mitigation initiatives. The specific prices which are adopted by a company are guided by estimates of the social cost of carbon, analyses of marginal abatement costs required to meet internal reduction targets, or benchmarking against the market price of carbon. Since 2012, Microsoft has used an internal carbon price for company decisions. This was increased to $15 per metric ton of carbon in 2019 [9]. Internal carbon pricing adoption has risen 89% since 2021, with nearly 1,800 companies worldwide taking this approach [3].

3.4 Voluntary and Cooperative Approaches

Voluntary and cooperative approaches are not mandated legally but are adopted by companies based on self-regulation and shared commitments. Table 3.11 provides a description and examples of different policy instruments using voluntary and cooperative approaches to drive climate action.

3.5 Information, Education and Capacity Development

These policies are aimed at increasing awareness through informing and educating consumers, which may lead to behavioral change. They can be broadly segregated into awareness and information, formal and informal education, and skills and capacity development. Table 3.12 provides a description and examples of different policy instruments using such approaches.

Table 3.11 Different policy instruments and examples of voluntary approaches

Instrument	Description	Examples
Voluntary pledges	Companies voluntarily commit to GHG emission reductions, energy efficiency or procurement of renewable energy	Science Based Targets initiative (SBTi) is a voluntary corporate climate initiative where companies commit to reducing GHG emissions by setting targets which are consistent with limiting warming to 1.5 °C [16] Oil and Gas Methane Partnership is an initiative where energy companies voluntarily agree on a measurement-based reporting framework, which improves the accuracy and transparency of methane emissions data from the oil and gas sector [14] Japan's Keidanren Voluntary Action Plan is an industry-wide commitment to improve energy efficiency [7]
Corporate social responsibility (CSR) and environmental, social and governance (ESG) initiatives	Companies embed climate goals into broader corporate and environmental performance strategies. Many companies have declared net-zero corporate goals	Microsoft has a goal to become carbon negative by 2030. It has a $1 billion climate innovation fund to accelerate technology development and deployment of new climate innovations Google ESG initiatives include procurement of 100% RE (equal to annual electricity consumption) for the past 8 years IKEA invests CSR funds in sustainable forestry
Community-based and local cooperative initiatives	Local actors voluntarily implement mitigation or adaptation projects which empower local communities and builds ownership of climate goals	Energy cooperatives in Europe pool local investment and channel them into RE and efficiency projects through Community Energy Financing Schemes (CEFS) Community forestry initiatives in Nepal combat deforestation and manage land sustainably

(continued)

Table 3.11 (continued)

Instrument	Description	Examples
Negotiated agreements	Long-term voluntary or semi-voluntary commitments between governments and industries to achieve emission reductions or energy efficiency improvements. These agreements may include a clause where companies receiving additional benefits may lose them if they do not achieve the commitments	The EU auto industry has an agreement between the European Automobile Manufacturers' Association (ACEA) and the European Commission to cut average CO_2 emissions by 25% from new cars between 1998 and 2008 In the Netherlands, there were Long Term Agreements (LTAs) with industries to achieve a 30% improvement in energy efficiency from 2005 to 2020 and use of RE in companies
Voluntary reporting networks	These are frameworks for exchanging data, experiences, and methodologies which build capacity for climate action without any obligations	Global Covenant of Mayors for climate and energy is a global alliance where cities voluntarily report GHG inventories and climate actions

3.6 Policy Packages and Instrument Mixes

A policy package is a set of coordinated policies which are bundled together to achieve strategic goals. Policy packages can be used to attain climate goals while simultaneously achieving energy security, economic competitiveness, and sustainable development.

3.6.1 Instrument Mixes and Complementarity

A climate policy package comprises multiple, mutually supportive and coherent policy instruments. Such policy packages include a combination of regulatory, economic and voluntary instruments which work together. Policy packages are required because individual policies alone are insufficient to address the scale and complexity of climate change. A bouquet of policies is therefore necessary as different policy instruments address different stakeholders and types of market failures. A well-designed policy package addresses different emission reduction pathways simultaneously.

Climate policies that implement multiple instruments together tend to be more effective than individual policy interventions. This is because bundling policies

Table 3.12 Different policy instruments and examples of information, education and capacity development

Instrument	Description	Examples
Public awareness and information campaigns	These awareness campaigns are designed and implemented by local or national authorities to inform the public on climate change and its impacts. By building public awareness, they attempt to influence behavioral and lifestyle changes	UNFCCC Action for Climate Empowerment (ACE) program aims to empower the public to engage in climate action through increasing public awareness and access to information 'Act on CO_2' campaign launched in late 2000s in the UK, promoted household energy savings and low-carbon lifestyles India's National Environment Awareness Campaign (NEAC) engaged schools and NGOs in localized mass awareness campaigns to promote environmental sustainability
Public reporting and performance benchmarking	These instruments increase transparency by ranking or comparing the performance of institutions or entities on climate action	'Climate Action Tracker' and 'Net Zero Tracker' track climate action and assess national or corporate commitments against the Paris climate agreement targets Dow Jones sustainability index is used in the stock market to track companies' sustainability performance, to benchmark and identify leading companies in sustainable business practices

(continued)

Table 3.12 (continued)

Instrument	Description	Examples
Voluntary information disclosure	Provide voluntary information disclosure to consumers and investors, which increases transparency and provides higher recognition due to better climate performance	CDP (Carbon Disclosure Project) is a voluntary corporate disclosure of emissions and climate risks by corporations, cities and governments, which enhances the credibility of the organization's environmental strategy and increases investor confidence UN Global Compact is the world's largest corporate sustainability initiative to adopt sustainable policies, and to report on their implementation ISO 14064 certification for GHG accounting enables transparent reporting of carbon footprint
Product labeling and certification	Provide consumers with visible information about the environmental impact, carbon footprint and relative energy consumption of a product or a service. Products with low carbon footprint are preferred by an informed consumer resulting in higher sales	Green building certifications such as LEED and BREEAM boost reputation and impact sales Energy Star (USA) and EU Energy label systems for appliances on a standardized A-G scale helps consumers compare energy performance of products Carbon Trust Label (UK) shows the total carbon footprint of the product, expressed in kg CO_{2e} Forest Stewardship Council (FSC) certification guarantees that forest products come from responsibly managed forests

(continued)

Table 3.12 (continued)

Instrument	Description	Examples
Weather information and early warning systems	These provide real-time or predictive weather data for decision-making in agriculture, disaster management, and infrastructure planning for increasing resilience and climate risk reduction	WMO Global Framework for Climate Services (GFCS) supports countries in developing early warning systems and integrating monitoring tools and climate services into national systems and reporting frameworks Attributes of Recycled Content (ARC) group helps governments to improve their capacities to prepare and respond to extreme weather events and access financing India Meteorological Department's (IMD) Agromet advisory service delivers local weather and crop advisories to farmers helping them to mitigate weather-related risks
Climate data, research, and information portals	Provides open access to scientific data, climate projections, and climate policies, which enables citizens to access climate data	IPCC data distribution centre provides an accessible archive for climate, socio-economic and environmental data and scenarios used in the reports Earth Science Data Systems (ESDS) program provides open access to NASA's archive of Earth science data World Bank Climate Change Knowledge Portal provides climate-related data on historical and future climate, vulnerabilities, and impacts National GHG inventories are publicly accessible on the UNFCCC website

(continued)

Table 3.12 (continued)

Instrument	Description	Examples
Introduction of climate change curriculum in formal education	Embedding climate change into school, university, and vocational training curricula develops climate-aware citizens and a skilled, future-ready workforce	In 2020, Italy made it compulsory for schools to teach climate change and sustainable development UNESCO's Education for Sustainable Development (ESD) for 2030 program offers guidance and technical support to countries, and implements climate change education projects
Technical training programs for professionals	Targeted training for policymakers and professionals to strengthen technical ability to design and implement climate mitigation and adaptation programs	UN CC: Learn offers courses on climate change for skill enhancement and capacity development FAO's Climate-Smart Agriculture (CSA) program offers training for farmers on sustainability in agriculture while addressing the impacts of climate change ICLEI's urban resilience program offers training to develop adaptation and resilience strategies for local governments
Community-based education	Initiatives combine awareness and participatory learning to empower communities with localized knowledge and build capacity for collective action	Community-based adaptation training in Bangladesh and Pacific Islands to build local capacity to respond to climate change Non-governmental organization (NGO) led programs by *Practical Action* on sustainable energy access, waste segregation, and water conservation for rural communities *Citizen Science Global Partnership* runs an initiative which uses citizen-collected data to monitor and manage air quality

(continued)

Table 3.12 (continued)

Instrument	Description	Examples
Knowledge sharing and exchanging best practices	Voluntary or government-supported networks for exchanging data, tools, and experiences to accelerate adoption of proven climate solutions	Local Communities and Indigenous Peoples Platform (LCIPP) by UNFCCC integrates indigenous and traditional knowledge and practices into national and international climate policies and programs Climate-ADAPT platform by the European Environment Agency (EEA) provides information on current and future climate risks and vulnerabilities, tools and case studies for climate change adaptation across Europe International Civil Aviation Organization (ICAO) knowledge-sharing platform provides best practices on measures and strategies for reducing emissions from international aviation

together improves the cost-effectiveness of policies, which lead to measurable emissions reductions across different sectors of the economy. A policy package can also increase political acceptance of some policies as it can distribute the costs across the economy. For example, policies such as carbon taxation have low public acceptance when they are implemented in isolation. However, when these policies are combined with supportive instruments such as targeted subsidies or cash transfers from generated carbon tax revenue, providing rebates in health subsidies and public investment in clean infrastructure, they can improve the perception of fairness and make the overall climate policy package politically acceptable.

Climate policies also need to be complementary. Complementarity arises when one type of policy makes another policy more effective or less costly to implement. For example, a carbon tax provides a signal to high-emitting entities to lower emissions. Policies such as grants for R&D in low-carbon technologies can complement a carbon tax. Acting together, the carbon tax increases the demand for clean energy technologies and the grants for R&D help to route private investment for innovation. This results in lowering the cost of clean technology in the long term while simultaneously reducing emissions.

Policy complementarity requires an in-depth understanding of how policy instruments interact at different levels. Combining multiple climate policy instruments can create synergies that strengthen and reinforce the impact of other policies, increasing the overall effectiveness. For example, regulations to phase out diesel generators for power backup, when combined with subsidies for installation of rooftop solar panels, can lead to faster adoption of battery backup and lower emissions than either of the policies in isolation.

Policies may also demonstrate trade-offs which need to be managed through appropriate choice and design. For example, carbon taxes encourage power producers to buy power from cleaner energy sources, and in conjunction with producer subsidies for RE, it can lead to a higher share of low-carbon sources in the power mix. However, there is a trade-off between the economic burden on consumers due to an increase in the price of electricity. This could be addressed by providing targeted subsidies for consumers in low-income groups.

Effective policy mixes must therefore be carefully designed to balance trade-offs and increase complementarities, as it helps to achieve climate goals at a lower overall cost by leveraging the synergies between different policy instruments. Global evidence also suggests that combining voluntary measures with regulatory measures and financial incentives improves the effectiveness of climate policy packages.

A coherent policy package should be designed to respond to different types of market failures, which requires policy coherence and coordination. Policy coherence includes horizontal coherence within sectors, vertical coherence between municipalities, states, national and at the international level. Policy packages should be developed through systematic stakeholder engagement processes, involving local governmental bodies, ministries, agencies, industries, NGOs, academia and civil society. A coherent policy package ensures that different policy instruments work collectively to meet the end goals rather than at cross-purposes.

The sequencing of policies and their timing also plays a crucial role in increasing the impact of policies. Delaying climate action generally results in much higher overall economic costs and more difficult transitions. Early climate action, including investing in adaptation, can lead to benefits during the transition stage, increasing the payoffs. Policies can be more effective if introduced in stages. Early policies which form the initial core of a climate policy portfolio include information campaigns, grants, subsidies and regulatory instruments. This could be followed by incentives for R&D, voluntary agreements, and public procurement policies. More stringent policies that rely on carbon trading mechanisms should be introduced more gradually. Prior to implementing climate policies, it is important to prepare the enabling conditions. Policies for increasing awareness can lower public opposition to more stringent regulatory measures. Targeted subsidies can also assist in lowering the political and economic barriers to more stringent measures such as carbon taxes.

3.6.2 *Examples of Policy Instrument Mixes*

Germany's Energiewende, which is an ambitious plan to transition from fossil fuels to a RE-based economy, improving energy efficiency and reducing GHG emissions, is a good example of how policy instruments can be effectively used together. Germany's 'Renewable Energy Act' provided 20-year feed-in tariffs (FiTs) for wind and solar, which allowed banks to finance RE projects due to an assured future revenue stream. Utilities and grid operators were also required to purchase electricity generated from RE, which was granted priority grid access over electricity generated from fossil fuels. This lowered the risk of RE curtailment, leading to scaling up of renewables. This also led to the creation of several citizen-owned wind turbines and solar cooperatives. By 2012, over half of the RE installed capacity was owned by households and local cooperatives instead of utilities. This was one of the first examples of a decentralized ownership model for RE, which led to community support for RE as the economic benefits were distributed amongst the public, building political legitimacy for the clean energy transition. Energy efficiency improvements were coordinated across sectors through adoption of aligned policies. KfW (the German state-owned investment and development bank) provided loans for funding energy-efficiency improvements in buildings and industries. This was accompanied by implementation of regulations for energy efficiency and increasingly stringent electrical appliance standards, which gradually raised minimum energy performance levels and led to lower energy intensity across multiple sectors. Germany's Energiewende was successful in implementing a whole-of-system governance approach where ministries, local utilities and cooperatives aligned around a common goal through implementing a policy package where policies reinforced each other.

Norway's electric vehicle policy provides another evidence of how policy mixes can work in unison. Norway exempted EVs from sales taxes, registration tax, VAT and import taxes, which offered incentives for consumers to purchase an EV. Along with this, the government allowed EVs to access bus lanes, offered discounted ferry access, and reduced parking fees and tolls, which provided owners with additional savings of time and money. The government co-funded the deployment of fast charging infrastructure along the highways and municipalities also invested in public charging facilities. Long-term political commitment was provided through bipartisan support to the EV transition, ensuring continuity of policies. National authorities implemented fiscal policies while municipalities and counties implemented on-the-ground measures which ensured vertical policy coherence. By the end of 2025, over 95% of new car sales in Norway were fully electric, demonstrating that well-designed and coherent policy packages can rapidly lead to a sectoral transition.

Costa Rica's forest conservation program provides another successful example of a well-designed policy mix. Costa Rica offered payments for ecosystem services (PES) to landowners who received financial compensation for conserving forests, watersheds, and biodiversity. This economic incentive was supported with legal property rights and opportunities for ecotourism development, which created alternative livelihoods. Legal support was also provided through enacting regulations for forest protection and penalties for deforestation. Technical assistance and capacity building initiatives included training and management support for enabling participation of locals. As a result, Costa Rica's forest cover increased from around 20% in the 1980s to over 50% by the 2010s. This demonstrates the power of combining economic, regulatory, and informational policies for environmental conservation.

It is evident that an effective climate policy package is more than just a combination of individual policy instruments. It is a strategically designed package where all policy instruments collectively achieve ambitious climate goals. There is no 'one size fits all' policy package, and the optimal policy mix depends on national circumstances, where each policy is tuned to respond to a specific need.

3.7 Barriers to Policy Implementation

3.7.1 Political Economy

Stakeholders have varying and often competing interests. Nationalized and private fossil fuel companies, industries which have fossil fuels-based infrastructure, and governments of oil, gas and coal exporting countries act in their vested interest to delay the clean energy transition. They spend financial resources and exert political influence to resist changes, as a clean energy transition could reduce their revenue streams and profits. Lobbying expenditures often overshadow environmental advocacy budgets and politicking leads to weak regulations, which have delayed implementation timelines. Short-term electoral cycles are misaligned with the longer time frame which is required for implementing meaningful climate policies. This creates political challenges for incumbent governments and discourages adoption of stricter climate goals and unpopular policies, which may lead to strong political opposition.

Unequal distribution of costs and benefits is another challenge that needs to be addressed. Decarbonization policies may lead to job losses and disrupt revenue for coal-producing countries, which may be seen as an economic burden and is considered socially unacceptable. Regions which are rich in these natural resources may have to bear disproportionate costs of the clean energy transition, which creates political resistance. On the other hand, most benefits from climate action, such as improved air quality or lower impacts of climate change, are geographically diffused. This has led to the call for a 'just and equitable transition'.

3.7.2 Economic Competitiveness

Governments are concerned that stringent climate policies may increase costs in the short-term, leading to reduced industrial competitiveness and lower economic growth. High carbon taxes in a jurisdiction impact the production costs of energy-intensive and trade-exposed sectors like steel, cement, aluminum, and chemicals, making their products uncompetitive in the international market. To avoid additional costs, conglomerates may move production to countries with weaker regulations, leading to de-industrialization and carbon leakage, which is a genuine concern for global climate action. Measures like CBAM are being implemented to provide a level playing field to domestic industries, but concerns about economic competitiveness remain as the goods produced domestically have limited market potential outside the specific jurisdiction, due to higher prices.

Rapid decarbonization has a high transition cost, which could create a risk of stranded assets. A higher share of renewables, investment in clean hydrogen production and its transportation infrastructure create a risk of underutilization of existing fossil fuel assets such as coal-fired plants or reduced volume of oil and gas flow in pipelines, leading to lower revenue. Companies and financial institutions with sunk costs in fossil fuel assets resist the clean energy transition as their portfolio faces higher financial risks. Companies are also faced with the high cost of retraining workers in new technologies, which impacts their bottom lines. These concerns, coupled with slowing economic growth, may create policy reversals or slowdowns, which further amplify investment uncertainty. For economies in transition, competing development priorities, energy security concerns and the need for faster economic growth often push climate action to a lower priority.

3.7.3 Social Acceptance

While public awareness on climate change is increasing, low climate literacy continues to be a barrier for advancing climate action. The imperative of earning a living, especially in low-income countries or among low-income groups, makes it difficult to prioritize climate action. In certain countries, misinformation campaigns are used deliberately to create confusion and scientific knowledge and institutions are systematically dismantled. Behavioral barriers also emerge from time discordance as climate change is perceived as something which will happen beyond one's lifetime, which reduces the urgency to act.

Climate action is subdued on an individual level as it requires lifestyle changes such as dietary shifts, reduced air travel and lower comfort. There is a cultural and personal resistance to using public transport, segregating waste and lower use of resources, as some of these habits are deeply embedded in lifestyles. Regulatory policies such as technology bans may be perceived as interference with personal choice and autonomy. Climate policies could be perceived as unfair, as carbon pricing or taxes may disproportionately burden low-income households, as it increases energy prices. This creates social opposition as it does not adequately address distributional equity. Intergenerational equity concerns and unresolved issues related to historical emissions also create barriers to acting on climate change.

3.7.4 Technical and Institutional Capacity

Technical and institutional capacity for the design of polices and their implementation is often limited in developing countries. This may include technical expertise, administrative capacity, weak institutions, and inadequate MRV systems. Adoption of multiple standards, lack of coordinated methodologies and different regulations across jurisdictions create challenges for implementation. Lack of coordination amongst government departments and institutional fragmentation is a major barrier to policy adoption and implementation. Bureaucratic hurdles and political upmanship limit a coordinated approach to policy design. Multi-level governance challenges and lack of coordination between national, regional, and local authorities, often due to political differences, create implementation gaps, while poor horizontal coordination results in conflicting policies across sectors. Infrastructure gaps such as limited access to modern electricity grids and electric charging networks for transportation hinder ambitious climate action. Access to advanced technologies may be protected by intellectual property rights or high-cost barriers. Deployment and scaling up bottlenecks may include workforce skills gaps and supply chain constraints.

3.7.5 Policy Design Challenges

Policy design and selection of policy instruments is challenging and needs constant improvement based on evidence. Rapid changes in technology and behavioral responses to these changes complicate policy design. Choice of appropriate policy instruments such as taxes, regulations and incentives requires a detailed understanding of the interactions and is context-specific. Setting free emission allowances in a cap-and-trade system is difficult and may need to be balanced with political reality and competitiveness concerns. Discovery of an optimal carbon tax is challenging, as a low tax fails to drive behavioral change while a high tax may trigger political backlash. A one-size-fits-all approach may not be suitable, and sectoral heterogeneity and regional circumstances must be addressed in policy design.

Policy incoherence often creates competitive distortions. One example is untargeted fossil fuel subsidies, which often contradict climate policies. Temporal inconsistencies such as oil and gas subsidies in 2022 in the European Union during the COVID-19 pandemic led to delays in meeting climate objectives. Sometime well intended policies have unintended consequences. These include rebound effects where improvements in energy efficiency may lead to higher energy consumption due to behavioral changes in individual users. Biofuel blending mandates may lead to increased deforestation and higher food prices. Capital subsidies for solar rooftop systems may benefit wealthy homeowners disproportionately. These policies therefore require careful consideration before being implemented and must be reviewed regularly for effectiveness.

3.8 Measuring Policy Effectiveness

Measuring policy effectiveness is important as it allows governments to track the impact of existing climate policies. It also provides evidence for continuing policies or making suitable modifications to increase the effectiveness of policies. Measuring climate policy effectiveness requires data to quantitatively assess the impact of policies. Qualitative measures are also employed to evaluate the socio-economic impacts. Measuring policy effectiveness also strengthens transparency and creates accountability for policymakers. Based on the effectiveness of existing policies, governments decide future policy actions that maximize the positive impacts and minimize the unintended negative impacts. It is difficult to evaluate the effectiveness of individual policies as they interact with each other, and isolating the impact of a single policy is challenging.

3.8.1 Evaluation Frameworks

Evaluation frameworks provide a structured approach for assessing the impact of climate policies and measuring their effectiveness. Some commonly used evaluation methodologies are briefly described ahead.

3.8.1.1 Cost–Benefit Analysis

Cost–Benefit Analysis (CBA) involves comparing the total costs of implementing a climate policy against its overall benefits, which includes financial and non-financial impacts. To implement the CBA framework, it is important to identify and define the scope of the impacts. First, all costs such as subsidies, loss of revenue and administrative costs are calculated. Thereafter, all environmental and social impacts and co-benefits from implementing the polices such as avoided health costs from reduced air pollution and productivity improvements, are assigned a monetary value. Once the costs and benefits are quantified, a net present value assessment tool can be used to determine if the benefits of implementing a policy are greater than its costs.

For example, to evaluate the effectiveness of EV policies in Norway, the total cost of implementing the policy, which included subsidies, tax exemptions, lower tolls, and other expenditure such as public charging infrastructure, is evaluated. This is followed by assigning a monetary value to the societal benefits such as better air quality, additional jobs, improved public health and better energy security. It was concluded that the benefits exceeded the costs over the EV's lifetime and hence the policy was continued.

Some challenges of CBA include difficulty in quantifying and assigning a monetary value to environmental and social impacts, especially long-term impacts of climate change, as it may involve significant uncertainty. The method also does not address equity issues such as distribution of costs and benefits across different income groups.

3.8.1.2 Multi-criteria Analysis

Multi-criteria analysis (MCA) is a method to integrate social, environmental, economic, ease of implementation and other criteria that are difficult to quantify in monetary terms. It provides a structured framework to evaluate a policy against selected multiple objectives and can handle a mix of quantitative and qualitative data.

Before applying MCA, the scope of the policy must be defined. A set of alternative policies that need to be evaluated must be listed as different options and a set of criteria against which the performance of each option needs to be identified must be fixed. This may include impacts such as reduction of GHG emissions, financial costs, social aspects such as job creation, technical feasibility, administrative burden and other such criteria. The MCA process involves the judgement of stakeholders such as policymakers, researchers, and experts to identify the criteria and to assign weights to these evaluation criteria as per their relative importance. The performance or potential of each of the policy alternatives against the identified criteria is then scored using expert judgment. Individual scores allocated by experts are aggregated for each policy option using specialized techniques such as the analytical hierarchy process (AHP) and the relative ranking of each policy option emerges from the process. The MCA process is often followed by conducting a sensitivity analysis. This includes an analysis to identify how the relative ranking of policy options changes by varying the weights allocated to different criteria. This allows incorporation of uncertainty in data and expert judgment, thereby enhancing the transparency and acceptability of the results.

As an example, Rwanda used the MCA method to prioritize different climate adaptation options in National Adaptation Programmes of Action (NAPAs). Several policy options, including promotion of rain-fed agriculture, integrated water resource management, intensive agro animal husbandry, varieties of seed resistant to drought, and stocking and transformation of agricultural products were identified. The criteria selected for prioritizing adaptation options and their chosen weights were as follows: level of climate risk (0.30); poverty reduction potential (0.20); cost effectiveness (0.20); complementary with national plans (0.15); and synergy with national plans and multilateral environmental agreements (0.15). Based on the MCA process, integrated water resource management option emerged as the most preferred policy option and promotion of rain-fed agriculture was the least preferred policy option [2]. In this manner, the MCA process resulted in a priority list of adaptation projects based on which informed decisions could be taken to adopt relevant policies in a transparent manner.

3.8.1.3 Ex-Ante Versus Ex-Post Evaluation

Ex-ante and ex-post evaluations are complementary approaches for measuring climate policy effectiveness. Ex-ante evaluation is forward-looking and is conducted before a policy is implemented. It uses models and scenarios to estimate the likely impacts of a policy and helps inform policy design. Ex-post evaluation is backward-looking and is conducted during or after a policy has been implemented. It assesses the performance of an implemented policy using actual data such as emissions to determine the impact of the policy through statistical and econometric methods that link observed emission reductions to the implemented policy. As ex-post evaluations are based on observed data, it provides empirical evidence of the effectiveness of the policy by undertaking a counterfactual analysis and comparing actual emissions against 'business-as-usual' scenarios.

As an example, the EU used an ex-ante impact assessment to analyze the potential effects of setting the EU's 2040 climate target before it was adopted. This helped inform the policymakers on the potential benefits and impacts of the climate legislation. An example of an ex-post evaluation is the analysis of the EU ETS as part of the assessment of EU climate policies. The assessment, which was a synthesis of 187 studies covering five sectors and about 450 estimates of policy impact, concluded that most policies effectively reduce emissions, with a median effect of lowering emissions by about 5% per year, but there was substantial variation across studies [12].

3.8.1.4 Key Performance Indicators

An easier way to measure policy effectiveness is by tracking quantifiable metrics over time. Some of the common key performance indicators (KPIs) used to track the impact of climate policies are GHG emissions over time (e.g., total emissions,

emissions intensity), aggregate energy consumption (e.g., total energy use) and share of non-carbon energy in the primary energy mix (percentage of renewables). Other compound KPIs include measurement of energy consumption per unit of output or service (e.g., building energy efficiency can be monitored by tracking the building's annual energy consumption per unit of area, expressed in kilowatt-hours per square meter per year). Other indirect KPIs could include tracking annual investment flows in renewable energy technologies, technology diffusion rates, jobs created in renewable sectors and data on public awareness of climate impacts using surveys, amongst others.

3.8.1.5 Cost-Effectiveness Analysis

Cost-effectiveness analysis is a method to compare different climate policy options for achieving a predefined goal, for example reducing GHG emissions by 50%. The method assesses different policies to evaluate the cost per ton of GHG emissions abated to identify the cheapest policy option. Different policy options can then be ranked based on marginal abatement cost curves (MACC). For example, the marginal abatement cost of energy efficiency measures such as improved building insulation is assessed to be far lower than carbon capture measures. Hence, a policy which incentivizes improvement in building energy efficiency is preferable over policies which provide incentives for setting up carbon capture plants.

3.8.1.6 Policy Stringency Scores

Policy stringency is the strength of the policy signal or the degree to which a policy imposes an explicit or implicit cost on environmentally harmful activities such as GHG emissions. Policies are scored based on the stringency of various policy instruments as they impose different costs on GHG emissions, with higher scores indicating more stringent policies. The OECD's Environmental Policy Stringency (EPS) index is a tool that scores policies on a 0–6 scale, where 6 denotes most stringent policies [8]. Scores are allocated to various policy instruments, such as carbon taxes, emissions standards, feed-in tariffs, subsidies and R&D support. The EPS index aggregates the information on selected environmental policies using predefined weights to create a composite measure of the country's climate policy stringency, which can be used for international comparison. The OECD's Climate Actions and Policies Measurement Framework (CAPMF), which is an internationally harmonized climate mitigation policy database, uses a policy stringency index to track the evolution of climate action for over 50 countries [13]. Studies have shown that there is a correlation between higher policy stringency scores and lower GHG emissions, which indicates that it can be used to measure climate policy effectiveness.

3.8.2 Attributing Emission Reductions

Establishing causality and attributing emission reductions to specific climate policies is important for isolating climate policy impacts from confounding factors. A counterfactual analysis approach is often used for attribution studies. This approach uses different methods to estimate what GHG emissions would have been in the absence of the policy and to isolate the policy's impact from factors like economic growth, population changes, and technological innovation.

3.8.2.1 Difference-in-Differences

The Difference-in-Differences (DiD) analysis compares the change in an outcome variable such as GHG emissions over time for a treatment group that implemented the policy, versus a control group that was not subjected to the policy. This helps to isolate the policy's impact by calculating the difference in outcomes between the two groups before and after the policy was implemented. This neutralizes the impact of unobservable factors that affect both groups equally and allows for the policy evaluation using observed real-world data.

For example, this method can be used to evaluate the impact of carbon tax on GHG emission reduction. The treatment group will be a state where the carbon tax policy was implemented, while the control group will a similarly sized state that did not implement the policy. The change in GHG emissions for the treatment and the control group before and after the policy implementation is calculated. The second difference (control group's change) is subtracted from the first difference (treatment group's change) to obtain an estimate of the policy's causal effect. This eliminates the impact of any common factors such as global economic downturns on both groups.

3.8.2.2 Synthetic Control Method

Supplementary Cementitious Material (SCM) uses multiple control groups (e.g., states or countries that have not implemented the climate policy) to create a synthetic control unit using a weighted average. The synthetic control unit serves as the counterfactual and represents the hypothetical outcome for the treated unit if the policy had not been implemented. The climate policy's impact on the treatment unit is then compared with the outcome of the synthetic control unit in the post-policy period.

As an example, SCM was used to determine the impact of a pilot ETS program on improving air quality, measured by the Air Quality Index (AQI) in China [17]. A synthetic treatment unit was constructed as a weighted average of the nine cities in Zhejiang Province. Weights were assigned to the control cities based on certain characteristics like historical AQI levels, meteorological data and green area per capita to create a synthetic Zhejiang city. The observed post-intervention AQI of the treated cities was compared with the synthetic Zhejiang city (the counterfactual scenario) to estimate the policy's causal impact. The study concluded that there was a significant improvement in air quality during the period of implementation of the short-term regulations, but the impact was short-lived and the environmental benefits disappeared after the specific regulations were stopped.

3.8.2.3 Econometric Models

Certain statistical methods, such as fixed effects models, can control for unobservable, time-invariant factors. Method of moments quantile regression (MMQR) and computable general equilibrium (CGE) models can be used to isolate climate policy's impact on GHG emissions and to control for multiple variables such as GDP growth and population changes. Challenges in attributing emissions reductions to one specific policy include establishing a credible baseline by creating a counterfactual scenario, assessing the impact of interconnected drivers, data availability and quality and the time lag in observing the impact of climate policies. While most climate policies lead to a reduction in emissions, their effectiveness varies significantly. Hence, it is important to undertake climate policy evaluation for designing future policies.

3.9 Conclusion

Climate policies serve as a critical tool to deliver positive climate outcomes and catalyze long-term investment. Implementation of suitable policies can drive action and can effectively lower GHG emissions in a cost-effective manner. Effective climate action requires clear climate goals which are aligned with the Paris Agreement and accompanying transparency and accountability mechanisms. Addressing climate challenges requires comprehensive and integrated policy frameworks as the impact of policies transcends sectoral boundaries. Proven interconnections between energy, environment, climate, economy and society make it imperative to address these dimensions in climate policy. Implementation requires institutional and enforcement capacity along with strong political will to achieve high compliance rates. Policy evaluation can reveal the implementation gap between intended policy impact and the actual policy outcomes, which need to be addressed through adaptive management. Design and successful implementation of policies is an art as well as a science, and both these aspects are essential for policy effectiveness.

References

1. Department of Finance Canada (2021) Fuel charge rates for listed provinces and territories for 2023–2030. https://www.canada.ca/en/department-finance/news/2021/12/fuel-charge-rates-for-listed-provinces-and-territories-for-2023-to-2030.html. Accessed 25 Apr 2025

2. Dixit A, McGray H (2013) Analyzing climate change adaptation options using multi-criteria analysis. World Resources Institute/Tetra Tech ARD, for the United States Agency for International Development. https://www4.unfccc.int/sites/NAPC/Country%20Documents/General/Multi-Criteria%20Analysis_CLEARED.pdf. Accessed 28 Apr 2025
3. edie (no date) From Microsoft to Klarna: how businesses are using internal carbon pricing on road to net-zero. https://www.edie.net/from-microsoft-to-klarna-how-businesses-are-using-internal-carbon-pricing-on-road-to-net-zero/. Accessed 26 Apr 2025
4. European Commission (no date) Carbon border adjustment mechanism. Taxation and Customs Union. https://taxation-customs.ec.europa.eu/carbon-border-adjustment-mechanism_en. Accessed 25 Apr 2025
5. HM Treasury (2025) Factsheet: carbon border adjustment mechanism. GOV.UK. https://www.gov.uk/government/publications/factsheet-carbon-border-adjustment-mechanism-cbam/factsheet-carbon-border-adjustment-mechanism. Accessed 25 Apr 2025
6. Intergovernmental Panel on Climate Change (2022) Climate change 2022: mitigation of climate change. In: Contribution of Working Group III to the sixth assessment report of the intergovernmental panel on climate change. Cambridge University Press, Cambridge. [Cited within OECD (2024) The IFCMA's Climate Policy Database: Policy Instruments Typology and Data Structure. https://www.oecd.org/content/dam/oecd/en/publications/reports/2024/12/the-ifcma-s-climate-policy-database_91def4be/68529f35-en.pdf. Accessed 24 Apr 2025
7. Keidanren (Japan Business Federation) (1997) Outline of the Keidanren voluntary action plan on the environment. https://www.keidanren.or.jp/english/policy/pol058/outline.html. Accessed 27 Apr 2025
8. Koźluk T, Garsous G (2016) How stringent are environmental policies? OECD policy perspectives. OECD Publishing, Paris. https://www.oecd.org/content/dam/oecd/en/topics/policy-sub-issues/economic-policies-to-foster-green-growth/how-stringent-are-environmental-policies.pdf. Accessed 30 Apr 2025
9. Microsoft (2019) We're increasing our carbon fee as we double down on sustainability. Microsoft on the issues. https://blogs.microsoft.com/on-the-issues/2019/04/15/were-increasing-our-carbon-fee-as-we-double-down-on-sustainability/. Accessed 26 Apr 2025
10. Nachtigall D, Lutz L, Cárdenas Rodríguez M, Haščič I, Pizarro R (2022) The climate actions and policies measurement framework: a structured and harmonised climate policy database to monitor countries' mitigation action. OECD Environment Working Papers, No. 203. OECD Publishing, Paris. https://www.oecd.org/content/dam/oecd/en/publications/reports/2022/11/the-climate-actions-and-policies-measurement-framework_c473da5a/2caa60ce-en.pdf. Accessed 24 Apr 2025
11. OECD (2024) The IFCMA's climate policy database: policy instruments typology and data structure. In: Inclusive forum on carbon mitigation approaches papers, No. 5. OECD Publishing, Paris. https://www.oecd.org/content/dam/oecd/en/publications/reports/2024/12/the-ifcma-s-climate-policy-database_91def4be/68529f35-en.pdf. Accessed 24 Apr 2025
12. OECD (2025) The effects of climate policies on emissions: evidence from a comprehensive and systematic review of the ex-post empirical literature. In: Inclusive forum on carbon mitigation approaches papers, No. 7. OECD Publishing, Paris. https://www.oecd.org/content/dam/oecd/en/publications/reports/2025/05/the-effects-of-climate-policies-on-emissions_e8c0a91b/ec1f7f0d-en.pdf. Accessed 30 Apr 2026
13. OECD (2025) The climate actions and policies measurement framework (CAPMF). Data Explainer. https://www.oecd.org/en/data/insights/data-explainers/2025/04/the-climate-actions-and-policies-measurement-framework-capmf.html. Accessed 30 Apr 2026
14. Oil & Gas Methane Partnership 2.0 (no date) Homepage. United Nations Environment Programme. https://www.ogmpartnership.org/. Accessed 27 Apr 2025
15. Prahl A, Hofmann E (2016) Discussion of climate policy instrument types applied in the EU. Climate Policy Info Hub. https://climatepolicyinfohub.eu/discussion-climate-policy-instrument-types-applied-eu. Accessed 24 Apr 2025
16. Science Based Targets initiative (no date) About us: what we do. https://sciencebasedtargets.org/about-us. Accessed 27 Apr 2025

17. Wang C, Yan J, Li J (2024) Does China's carbon emission trading scheme reduce CO_2 emissions? Comprehensive evaluation from synthetic control method using lasso. Energy Reports. https://www.sciencedirect.com/science/article/pii/S235248472400547X. Accessed 30 Apr 2025
18. World Bank (no date) Carbon pricing dashboard: price. https://carbonpricingdashboard.worldbank.org/compliance/price. Accessed 24 June 2026
19. World Bank (no date) Carbon pricing dashboard: revenue. https://carbonpricingdashboard.worldbank.org/compliance/revenue. Accessed 24 June 2026
20. World Population Review (2026) Carbon tax countries 2026. https://worldpopulationreview.com/country-rankings/carbon-tax-countries. Accessed 24 June 2026

Chapter 4
Bridging the Climate Financing Gap

Abstract This chapter describes the critical role of climate finance for advancing climate action. It quantifies the climate finance commitments and the modeled climate finance need to identify the financing gap. Key sources of climate finance such as multilateral and bilateral public funding, national, private and philanthropic funding are described. Commonly used financial instruments for climate finance are presented. The chapter identifies the barriers to climate financing, such as investment risk, market, structural, capacity and policy barriers, and suggests key actions for bridging the climate finance gap. Successful models and examples for mobilization of climate finance are presented to bridge the climate finance gap.

Keywords Climate finance · Multilateral funds · Bilateral funds · Private funds · Financial instruments · Multilateral development banks

4.1 Introduction

The increasing impacts of climate change globally require significant financial resources to implement mitigation, adaptation and resilience strategies. Finance is a critical facilitator for climate action, and without adequate financing, climate impacts will continue to intensify, disrupting sustainable development. However, there is a large gap between funding required for effective climate action and the current levels of financial commitment, creating a climate financing gap, which is estimated at trillions of dollars annually. Quantifying the climate financing gap is essential to understand the scale of investment needed to address climate change. Closing this gap requires commitment, capital, mechanisms, policies, and collaborative efforts from governments and the private sector. Funds must also be allocated equitably and effectively to countries that are most vulnerable to climate change and to marginalized communities that are most impacted in underdeveloped economies.

This chapter describes the role of climate finance for enabling climate action. Section 4.1 commences with defining climate finance and identifying the climate finance gap. Section 4.2 describes the sources of climate finance such as multilateral

K. Narula, *The Climate Catalyst*, Advances in Sustainability Science and Technology,
https://doi.org/10.1007/978-981-92-5213-8_4

and bilateral public funding, national, private and philanthropic funding. Section 4.3 characterizes the instruments for climate finance, including debt, equity, grants and other commonly used financial instruments. Section 4.4 identifies the barriers to climate financing, such as investment risk, market, structural, capacity and policy barriers. Section 4.5 suggests key actions for bridging the climate finance gap. The last section presents successful models and examples for mobilization of climate finance, including capital markets, just energy transition partnerships, green sukuk and orange capital instruments, followed by the concluding section.

4.1.1 Defining Climate Finance

Climate finance refers to local, national or transnational financing which is sourced from public, private and other sources that seeks to support actions to address climate change [84]. Mitigation, adaptation, building resilience and compensating for loss and damage are some of the activities funded through climate finance. Mitigation activities aim to reduce greenhouse gas (GHG) emissions and enhance their sinks, while adaptation aims to reduce vulnerability to negative climate change impacts. Building resilience includes activities for maintaining and increasing the capacity of social, economic and natural ecosystems to cope with a hazardous event and often includes both mitigation and adaptation measures. Loss and damage financing compensates for unavoidable losses which occur from irreversible impacts of climate change. While loss and damage funds could be considered a part of the broader climate finance landscape, it is distinct from mitigation or adaptation financing.

4.1.2 Current Climate Financing

It is estimated that global climate finance flows were around US$1.9 trillion per year in 2023 and early reports indicate that climate finance exceeded USD 2 trillion for the first time in 2024 [16]. The growth in annual global climate finance flows from 2018 to 2023 is shown in Fig. 4.1. As a comparison, the global annual military expenditure in 2024 was US$2.7 trillion per year and the global fossil fuel subsidies in 2024 were US$7.4 trillion per year [53]. Further, in 2025, climate finance accounted for only about 1.5% of the share of the global economy, estimated at US$126 trillion [54].

Between 2018 and 2023, mitigation finance more than doubled, increasing from US$757 billion to US$1.78 trillion. Adaptation received US$65 billion and financing for dual objectives was about US$58 billion [16]. In 2023, over 75% of finance for climate mitigation was directed to energy systems and transport, driven by solar PV, wind power, and EVs, while the balance 25% was routed towards buildings, infrastructure and industry.

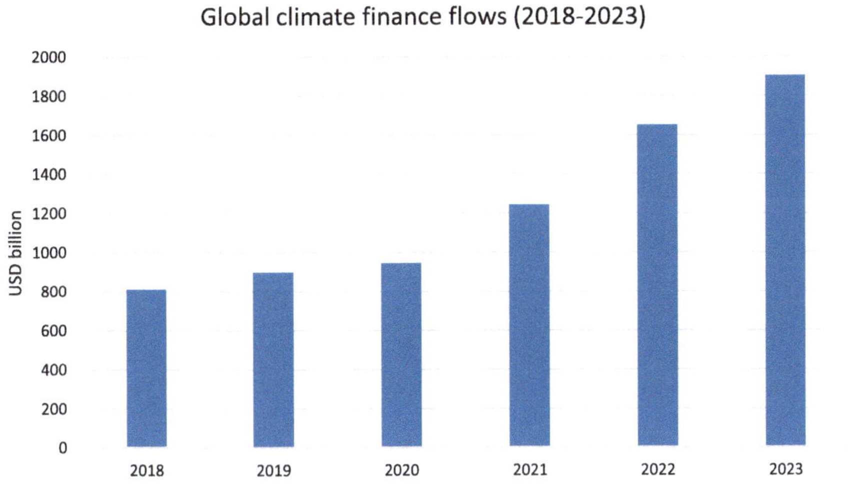

Fig. 4.1 Annual global climate finance flows (2018–2023)

4.1.3 Existing Commitment for Climate Financing

At COP15 (Copenhagen 2009), developed countries committed to mobilize US$100 billion annually by 2020 to support climate action in developing countries. This goal was met for the first time in 2022 with about US$116 billion mobilized in climate finance.

As shown in Fig. 4.2, it has been met continuously since 2022, and over US$136 billion was mobilized in 2024. Public climate finance from bilateral sources and multilateral public climate finance attributable to developed countries (including funds from multilateral development banks, multilateral climate funds and inflows to multilateral institutions) accounted for about three-quarters of the total in 2023 and 2024. Bilateral climate finance was close to US$44 billion, while multilateral public climate finance reached US$57.7 billion in 2024. Private finance mobilised by public climate finance (comparable data available from 2016 onwards) increased by US$7.6 billion in 2024. Climate-related export credits constitute a small share and contributed to US$4.6 billion in 2024 [67].

Over time, climate financing needs have evolved, and the US$100 billion climate financing goal is now considered grossly inadequate. The New Collective Quantified Goal (NCQG) on climate finance with a revised target was agreed at COP29 in November 2024. Countries agreed on 'core' financing, where developed nations agreed to take the lead, and an 'additional layer' encompassing private financing [81]. Parties agreed to:

(a) Triple finance to developing countries, from the previous goal of US$100 billion annually, to US$300 billion annually by 2035.

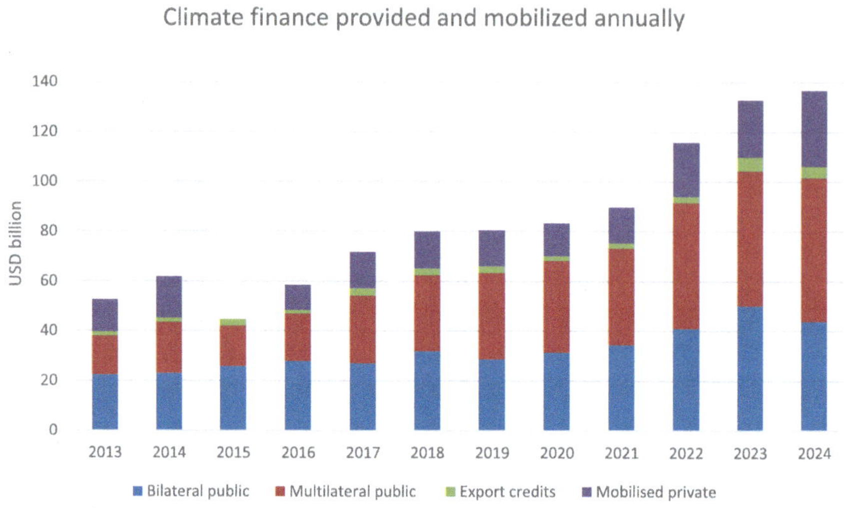

Fig. 4.2 Sources of climate finance provided and mobilized annually. *Data source* OECD [67]

(b) Secure efforts of all actors to work together to scale up finance to developing countries, from public and private sources, to the amount of US$1.3 trillion per year by 2035.

The '*Baku to Belém Roadmap to 1.3 T*', was published at the end of 2025, which aimed at scaling up climate finance to developing countries and sought to develop an action framework for mobilization of global financial flows [82]. It was also agreed to triple adaptation finance by 2035 (to $120 billion).

4.1.4 Estimates of Climate Finance Required

Despite the increasing climate finance commitments, the world faces an annual shortfall of several trillion dollars. A top-down needs assessment compiled from multiple data sources indicates that in an average scenario, annual climate finance flows must reach US$ 7.5 trillion per year till 2030, and over US$ 8.8 trillion per year between 2031 and 2050 [23]. The top-down climate finance needs are estimated using predictive models across different sectors to keep the average global temperature rise within 1.5 °C by the end of this century. Within mitigation, large investment gaps are observed in the energy and transport sectors, where an additional US$2.2 trillion and US$1.4 trillion per year will be required, respectively, till 2050.

In 2023, the public adaptation finance flows from developed country Parties to developing country Parties were about US$26 billion, which was lower than in 2022. The Glasgow Climate Pact goal of doubling international public adaptation finance from 2019 levels to around US$40 billion by 2025 is unlikely to be met if this trend continues. The cost of adaptation for developing countries by the year 2035 is estimated to be US$310–365 billion per year based on modeled costs and extrapolated needs expressed in Nationally Determined Contributions (NDCs) and National Adaptation Plans (NAPs). The adaptation finance gap is now US$284–339 billion per year until 2035 and the adaptation finance needs are 12–14 times more than the current adaptation finance flows [80].

Expanding these models to cover newer and increasing risks, a greater number of countries and more accurate estimates will further widen the estimated finance gap. Including second-order effects such as the impact of losses and damages for developing countries on developed economies through international and transboundary risk cascades will further widen the gap, as adaptation finance flows are not keeping pace with the financing needs.

4.1.5 Regional Disparities

Regional disparities in climate financing reveal stark inequities.

Figure 4.3 compares the share of global climate finance flows and socio-economic indicators in different country groups in 2022. Emerging Markets and Developing Economies (EMDEs), including China, Least Developed Countries (LDCs) and Small Island Developing States (SIDS), represented 87% of the global population, emitted 68% of CO_2 emissions and accounted for 42% of global Gross Domestic Product (GDP), but received 59% of climate finance. Within EMDEs, China alone received 42% of climate finance (CPI 2024). Countries most at risk, such as LDCs and SIDS, receive disproportionately low financial support, despite having contributed the least to global emissions. LDCs are among the most climate-vulnerable nations, but they receive only 2.7% of the global climate finance. 9 SIDS countries are among the top 25 countries at climate risk and are particularly vulnerable. They emit just 0.4% of CO_2 and receive only 0.5% of global climate finance. In contrast, advanced economies, including the US, Canada and Western Europe, while housing only 13% of the global population, receive 40% of climate finance. This clearly reveals a mismatch between the needs of developing countries and actual climate finance flows.

4.1.6 Impacts of Insufficient Climate Finance

Climate change has substantial economic consequences. According to estimates, extreme weather events, including heat waves, floods and wildfires, have cost the world economy US$2.8 trillion in damages from 2000 to 2019, which is over US$16 million per hour [94]. This economic loss undermines the ability of countries to achieve sustainable development goals and has a large impact on the environment,

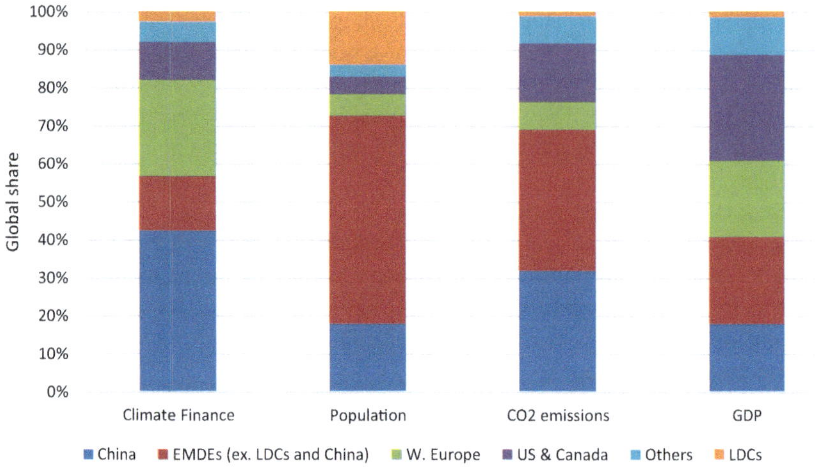

Fig. 4.3 Share of global climate finance and socio-economic indicators in different country groups in 2022

human development and economic growth. The climate crisis is not 'gender neutral'. Women are the most affected by climate change and migrants, Indigenous communities, elderly and disabled people are among the other groups that are frequently at risk from climate change. This amplifies existing socio-economic inequalities and creates compounding climate vulnerabilities, adding to the threats to the livelihoods, health and safety of these groups. Therefore, it's critical to view climate action and climate finance from a people-centered perspective that recognizes the varying effects of the climate crisis on various demographic groups.

Delaying further action on climate change has significant economic consequences, as the anticipated economic losses that could be averted by 2100 through achieving the 1.5 °C warming scenario are expected to be fivefold more than the climate funding required by 2050 to accomplish this goal (CPI 2024). Delayed climate action due to insufficient climate finance will lead to countries losing the gains they have achieved in building infrastructure and overall development and future expenses due to climate impacts will rise dramatically. In sum, the costs of doing nothing, as well as the harm and financial losses that climate change will bring to people, infrastructure, ecosystems, and incomes, are far higher than the costs of taking immediate action.

4.2 Sources of Climate Finance

Climate finance can be classified into public or private, national or international, bilateral or multilateral depending on the source of funds. International public sources for climate finance include funding provided from developed to developing countries through multilateral and bilateral sources. This is routed through international public institutions, development finance institutions, aid agencies and multilateral development banks. There is an established mechanism to provide financial resources to developing countries. The mechanism is accountable to the COP, which decides on its policies, programme priorities and eligibility criteria for funding. The Global Environment Facility (GEF) and the Green Climate Fund (GCF) are the designated operating entities of the financial mechanism [84]. Domestic public sources include domestic funds by national and sub-national governments, state-owned development finance institutions, and other state-owned entities. Some governments operate National Climate Funds (NCFs), which enable them to mobilize funds from a variety of domestic sources, pool them together and decide on their use, in accordance with national capacities.

4.2.1 Multilateral Public Funding

Some of the key dedicated multilateral climate funds include the following.

(a) Global Environment Facility

The Global Environment Facility (GEF), established in 1992, is a multilateral environmental fund for addressing environmental challenges and provides finance to developing countries for projects that generate global environmental benefits, including climate mitigation and adaptation projects. It serves as a financial mechanism for several key environmental conventions, including the UN Framework Convention on Climate Change (UNFCCC) and the Paris Agreement. Mitigation projects include reducing GHG emissions through renewable energy, energy efficiency, and land-use management, while adaptation projects aim to build resilient infrastructure for water management and agriculture.

The GEF also administers climate funds for adaptation, which target the unique needs of particularly vulnerable nations. The Special Climate Change Fund (SCCF) was established under the UNFCCC in 2001 and complements other funding mechanisms for the implementation of the Convention. It finances projects relating to adaptation; technology transfer and capacity building; energy, transport, industry, agriculture, forestry and waste management; and economic diversification [85]. The Least Developed Countries Fund (LDCF) was established to support a work program to assist LDCs to carry out, inter alia, the preparation and implementation of national adaptation program of action (NAPAs) [86].

The GEF operates on an indefinite mandate with a replenishment cycle of four years. GEF-8 covers the period from July 1, 2022, to June 30, 2026, with a total replenishment of US$5.35 billion. Climate change is one of the five identified focal areas, and approximately US$905 million was allocated from the GEF-8 cycle for climate change activities. GEF-9 cycle runs from July 1, 2026, to June 30, 2030, and donor countries have pledged an initial US$3.9 billion for this replenishment cycle [32]. Over the past three decades, the GEF has provided more than US$27 billion in financing, mainly as grants, and has mobilized US$155 billion for country-driven projects [31].

(b) Green Climate Fund

The Green Climate Fund (GCF) was established in 2010 at COP 16, Cancun, as an operating entity of the financial mechanism of the UNFCCC. It is the world's largest dedicated multilateral climate fund with a core mission to support developing countries to raise and realize their Nationally Determined Contributions (NDC) ambitions towards low-emissions, climate-resilient pathways [37]. The GCF operates through a network of accredited entities and country ownership ensures that funded activities align with national priorities and plans. It supports projects, programs, policies and other activities in developing country Parties using thematic funding windows. The GCF is mandated to invest 50% of its resources in mitigation and 50% in adaptation, in grant equivalent. A significant portion of adaptation funding is earmarked for the most vulnerable countries, including LDCs, SIDS, and the African States.

GCF's first replenishment period was from 2020 to 2023 with an announced funding of US$10 billion, and the fund invested in climate projects worth over US$40 billion (including co-financing) in more than 100 countries [38]. For the second replenishment period 2024–2027, 31 countries have pledged support with a total of US$12.8 billion [39]. GCF partners with over 200 accredited entities, banks, development finance institutions, UN agencies, and civil society organizations for project design and implementation. It structures financial support through a combination of grants, concessional debt, guarantees or equity instruments to leverage private investment for climate action in developing countries [37].

(c) Adaptation Fund

The Adaptation Fund was established in 2007 under the Kyoto Protocol of the UNFCCC and finances projects to help developing countries adapt to the adverse effects of climate change. It has served the Paris Agreement since 2019. The Fund supports localized projects focused on increasing resilience to climate risks such as extreme weather events, rising sea levels, and food and water insecurity. It offers various access modalities, including a 'Direct Access' mechanism, enabling national and regional implementing entities to directly access funding and manage all aspects of climate adaptation and resilience projects, from design through implementation to monitoring and evaluation [2]. This strengthens local expertise and capacity. Financed partly by a share of proceeds from Clean Development Mechanism (CDM) projects and voluntary contributions, the Adaptation Fund has committed over US$1.5 billion to adaptation and resilience projects since 2010, benefiting 217 projects in more than 110 countries [1]. The fund has received nearly US$135 million in new pledges at COP30 at the end of 2025 from 10 countries [3].

(d) Climate Investment Funds

The Climate Investment Funds (CIF), established in 2008, is a multilateral climate finance partnership that channels concessional finance through six Multilateral

Development Banks (MDBs) in developing and middle-income countries. These funds are administered by the World Bank Group, including the International Finance Corporation, the African Development Bank, the Asian Development Bank, the European Development Bank, and the Inter-American Development Bank [15]. By reducing financial barriers, the CIF aims to mobilize additional resources from the public and private sectors, accelerating the transition to a low-carbon and climate-resilient future.

CIF also helps the private sector to enter emerging markets and approximately 30% of its funding is invested in private-sector operations. CIF's combination of technical advisory and financial resources helps reduce investment barriers, test new business models, and de-risk new low-carbon, climate-smart markets, thus enabling the private sector to participate in emerging economies at a faster rate and on a greater scale. These funds are deployed via the MDBs to private-sector sponsors and companies through both national and regional investment plans, as well as dedicated funding mechanisms, such as CIF's Dedicated Private Sector Programs (DPSP) [15].

The CIF comprises two main funds. Clean Technology Fund (CTF), which provides large-scale financial resources for investing in clean technology projects in low- and middle-income countries. They contribute to the demonstration, deployment, and transfer of low-carbon technologies with significant potential for reducing long-term GHG emissions [15].

Strategic Climate Fund (SCF), is an overarching fund that provides financing for specific climate change challenges or sectoral responses targeted programs. These include the 'Forest Investment' program to address the drivers of deforestation and forest degradation and maintain important ecosystem services; the 'Pilot Program for Climate Resilience' to provide funding to build adaptation and resilience to climate change; the 'Scaling Up Renewable Energy in Low Income Countries' program to support deployment of renewable energy solutions; the 'Renewable Energy Integration' program which provides funding to address system-wide barriers to the integration RE into the grid; and the 'Nature, People and Climate' program to harness the potential of land resources and ecosystems in climate action [15].

CIF operates in more than 80 countries and has supported nearly 400 climate projects and programs with over US$12.5 billion pledged since its establishment. CIF funding has helped mobilize more than US$64 billion in co-financing from MDBs, governments, development finance institutions, and private investors as of 2024 [15].

(e) Multilateral Development Banks

Public financing is also provided by leading MDBs such as the World Bank Group, Asian Development Bank (ADB), African Development Bank (AfDB), European Bank for Reconstruction and Development (EBRD), and Inter-American Development Bank (IDB). MDBs have demonstrated commitment to climate financing over the years, reflecting the need for scaling-up action to address the climate crisis. Apart from providing direct financing, they also provide guarantees, technical assistance, policy support and mobilize private capital to assist developing and transition economies to address the impacts of climate change. Climate finance provided by MDBs reached US$137 billion globally in 2024, and they mobilized private finance worth US$33 billion. Of the total, US $85.1 billion was directed toward low- and middle-income economies, of which US $58.8 billion (69%) supported climate change mitigation, and US $26.3 billion (31%) was for climate change adaptation [7].

The World Bank Group has demonstrated a significant escalation in its climate finance, reaching a record US$50.8 billion in 2025, an increase over US$42.6 billion in the fiscal year 2024. Over the 2021–2025 Climate Change Action Plan (CCAP) period, the cumulative climate finance provided by the World Bank Group was US $190.3 billion. This comprises US$22.5 billion in financing for the International Bank for Reconstruction and Development (IBRD), and US$16.7 billion for the International Development Association (IDA) to support middle-income and low-income countries, respectively. Furthermore, the International Finance Corporation (IFC), the World Bank Group's private sector arm, contributed US$8.1 billion towards climate-related projects, while the Multilateral Investment Guarantee Agency (MIGA) provided US$3.5 billion in guarantees to encourage climate-friendly investments [93]. The funding provided is in line with the ambition to increase climate finance to 45% of its total lending portfolio by fiscal year 2025, with a growing emphasis on adaptation measures such as rising sea levels, extreme weather events, and water scarcity.

The ADB is also playing a crucial role in the Asia–Pacific region. In 2025, ADB committed US$13.5 billion in climate finance, representing 51% of its total annual committed financing, an increase over US$12.3 billion allocated for climate finance in 2024 [6]. The bank aims to deliver over US$100 billion in cumulative climate finance between 2019 and 2030. From 2019 to 2025, it had already reached US$55.4 billion towards this commitment, supporting projects in renewable energy, energy efficiency, sustainable transport, climate-resilient infrastructure, and sustainable agriculture [7]. The ADB also actively works to mobilize private sector investment in climate projects through various financial instruments and partnerships.

The AfDB is prioritizing climate action across the African continent. In 2024, AfDB allocated US $5.5 billion to climate finance, which was 49% of its portfolio, with 56% for adaptation and 44% for mitigation projects. The Bank is also on track to meet its USD 25 billion climate finance target for 2020–2025, having reached $19.5 billion (78%) by the end of 2024 [4]. The bank's focus areas include supporting climate-resilient agriculture, expanding access to clean energy, and investing in climate-resilient infrastructure.

The EBRD, operating in transition economies across Europe, Central Asia and North Africa, achieved a record investment of €16.8 billion in 2025 [28]. 56% (€9.4 billion) of this investment was dedicated to green projects, demonstrating the EBRD's strong commitment to fostering a low-carbon and climate-resilient future as a part of its portfolio. Furthermore, the EBRD mobilized €26.8 billion in total finance, indicating its effectiveness in attracting co-financing from other sources for its green initiatives.

The IDB and its private sector arm, IDB Invest, are important players in Latin America and the Caribbean. In 2023, they collectively dedicated US$7.5 billion to climate financing. The IDB Group aims to significantly scale up its climate finance, targeting a combined US$11.3 billion annually by 2030 [46]. A key priority for the IDB is supporting climate change adaptation with a goal of providing US$25 billion in adaptation finance between 2024 and 2030 [47].

As seen above, MDBs are demonstrating a growing commitment to climate financing. Through increased direct funding, strategic prioritization of climate-related projects, and mobilizing private capital, they play an important role that is crucial for funding projects to address climate change.

4.2.2 Bilateral Public Funding

Bilateral public climate finance is an important pillar of international public financing for addressing climate change. Unlike multilateral climate finance, which flows through international organizations with multiple member countries, bilateral public funding refers to the financial resources provided directly by a government or a public institution of the donor country to the recipient country. The funds originate from the public budget and are channeled to the government, national institution or specific projects for climate change-related activities within the recipient country.

Climate finance is emerging as an important thematic priority of the existing Official Development Assistance (ODA) from developed countries. Bilateral public funding for climate advances foreign policy objectives, strategic partnerships, commercial interests and soft power and contributes to the donor country's development cooperation agenda. Bilaterally funded projects have greater flexibility in tailoring financial instruments and project modalities. These can address specific needs and circumstances of the recipient country, which allows for the development of context-specific solutions. Several developed countries provide climate financing through their development agencies. Some of these countries have also established specific climate initiatives, which operate within their broader bilateral development cooperation frameworks.

(a) Germany: Germany is a large provider of bilateral climate finance, and funding is supported by the Federal Ministry for Economic Cooperation and Development (English translation) (BMZ) and the Federal Ministry for the Environment, Nature Conservation, Nuclear Safety and Consumer Protection (BMUV). Key agencies implementing climate projects include the Deutsche Gesellschaft für Internationale Zusammenarbeit (GIZ) for technical cooperation [25] and KfW Development Bank for financial cooperation [60]. Germany's International Climate Initiative (IKI), founded in 2008, also promotes climate action in the Global South [50]. The IKI supports countries in implementing and developing ambitious climate targets, adaptation, and biodiversity conservation projects in developing and emerging economies [51].
(b) United Kingdom: The UK's International Climate Finance (ICF) helps developing countries to tackle climate change and channels bilateral climate finance

through the Department for Energy Security and Net Zero (DESNZ) and the Foreign, Commonwealth & Development Office (FCDO). The ICF supports a wide range of mitigation and adaptation activities, with a strong focus on nature. The UK is allocating around £6 billion of ODA for climate over the next three financial years [44].

(c) Japan: The Japan International Cooperation Agency (JICA) provides financial and technical support for climate change mitigation and adaptation projects [40]. The Japan Bank for International Cooperation (JBIC) is a financial institution which conducts lending, investment and guarantee operations while complementing the private sector financial institutions. It provides financing for large-scale infrastructure and industrial projects with climate benefits [57].

(d) France: The Agence Française de Développement (AFD) is one of the world's leading international climate finance institutions and the principal French implementing agency for development projects. Since 2005, climate has been a core priority for AFD, and it has invested more than €33 billion in climate-related financing since 2017 [5]. France also supports climate action through the French Facility for Global Environment (English translation) (FFEM), which works on environmental protection and sustainable socio-economic development in developing countries through financing climate related projects [29].

(e) United States: The U.S. Agency for International Development (USAID) was an important channel for U.S. bilateral climate finance but was officially closed in the first half of 2025 [24]. The Development Finance Corporation (DFC) is the international investment arm of the U.S. government, which mobilizes private capital to advance U.S. foreign policy and economic development. It provides financing for critical minerals and energy projects apart from strategic sectors [77].

(f) Canada: Canada's bilateral programs are managed by Global Affairs Canada and Environment and Climate Change Canada, which channel bilateral climate finance to support projects in developing countries [34]. Canada's international climate finance program was established in 2010 and focuses on clean energy transition and coal phase-out, climate-smart agriculture and food systems, nature-based solutions and biodiversity, and climate governance [33]. Canada contributed C$2.65 billion from 2015–2016 to 2020–2021 and C$5.3 billion from 2021–2022 to 2025–2026 for climate finance [35].

(g) Norway: Norway's International Climate and Forest Initiative (NICFI) is an example of a targeted bilateral initiative focused on protecting the world's tropical rainforests in developing countries [65]. Norway also provides broader climate finance through the Norwegian Agency for Development Cooperation (NORAD) and the Ministry of Foreign Affairs. The Nordic Development Fund (NDF) contributed €729 million of climate finance in 2025 with projects in over 30 countries [64].

(h) Switzerland: Switzerland supports international climate action through bilateral cooperation projects. The Swiss Agency for Development and Cooperation (SDC) and the State Secretariat for Economic Affairs (SECO) are implementing agencies that provide bilateral climate finance, focusing on climate change mitigation and adaptation [75].
(i) European Union Member States: Besides Germany and France, other EU member states like Sweden, the Netherlands, Denmark and others provide significant bilateral climate finance through their respective development agencies and ministries. The EU also channels climate finance bilaterally through the Global Climate Change Alliance + (GCCA+). The GCCA works through the European Commission's established channels for political dialogue and cooperation and has an orientation towards supporting the Paris Agreement [11].

4.2.3 National and Subnational Public Climate Finance

Several countries have national and subnational public climate funds for accelerating climate action. Sources for these funds are from government budgets, often routed through national development finance institutions, state-owned financing institutions and state-owned entities. Some examples are listed ahead.

(a) India: National Adaptation Fund for Climate Change (NAFCC)

The National Adaptation Fund for Climate Change (NAFCC) was established in India in 2015 to support climate adaptation projects. It is managed by the Ministry of Environment, Forest and Climate Change (MoEFCC) and finances projects which are aligned with the priorities outlined in the State Action Plans on Climate Change (SAPCCs). The NAFCC aims to address adaptation measures in areas such as agriculture, horticulture, agro-forestry, environment, allied activities, water, forestry, urban, coastal and low-lying systems, disaster management, human health, marine systems, tourism, habitat sector and other rural livelihood sectors [61]. It operates through a project-based approach, supporting adaptation activities in the States and Union Territories of India.

(b) Indonesia: Indonesia Climate Change Trust Fund (ICCTF)

The Indonesia Climate Change Trust Fund (ICCTF) is a multi-donor fund hosted by the Indonesian government and aims to develop a low-carbon economy and greater resilience to climate change [12]. It pools financial resources from various national and international donors to finance climate related projects. Local governments can access funding via the ICCTF for financing mitigation and adaptation activities. The ICCTF operates through thematic windows, supporting initiatives in areas such as land-based mitigation, adaptation and resilience, energy and marine ecosystems.

(c) Bangladesh: Bangladesh Climate Change Trust Fund (BCCTF)

The Bangladesh Climate Change Trust Fund (BCCTF) was established by the Government of Bangladesh in 2010 for institutionalizing national climate finance [58]. It is funded through Bangladesh's own budgetary allocations and is managed by the Bangladesh Climate Change Trust (BCCT) and the government. The fund supports a range of projects focused on adaptation, mitigation, and climate change research and climate awareness.

(d) Philippines: People's Survival Fund (PSF)

The People's Survival Fund (PSF) was created by the Philippine government to implement climate change adaptation projects at the local level that will better equip vulnerable communities to deal with the impacts of climate change [70]. It prioritizes projects that address the impacts of climate change, such as extreme weather events, sea-level rise, and water scarcity. The PSF can finance activities ranging from infrastructure development and ecosystem-based adaptation to capacity building and technology transfer. The fund supplements the financial resources allocated by government agencies and local government units for climate-change-related programs and projects [69].

(e) Rwanda: Green Fund (FONERWA)

The Rwanda Green Fund, officially known as the National Fund for Environment and Climate Change (FONERWA), is one of the first national environment and climate change investment funds in Africa [87]. It mobilizes and channels domestic and international finance to support Rwanda's green growth agenda and climate resilience. Financing can be accessed by several actors in the national ecosystem through investment products such as grants, innovation investments, and credit lines. FONERWA finances a diverse portfolio of projects and programs aligned with Rwanda's climate action plans and sustainable development goals, covering areas such as sustainable agriculture, renewable energy, water management, and biodiversity conservation [71].

(f) South Africa: Green Fund

South Africa's Green Fund is a national fund managed by the Development Bank of Southern Africa (DBSA). It aims to catalyze investment in green economy initiatives that contribute to climate change mitigation and adaptation [26]. The Green Fund provides financial support targeting projects in water, energy, transport, and information and communication technology sectors [41]. It plays a crucial role in mobilizing both public and private sector finance towards South Africa's transition to a low-carbon and climate-resilient economy.

4.2.4 Private Funding

Private climate finance refers to investments made by project developers, commercial financial institutions, corporations, NGOs, and households for climate-related projects. Private finance needs to be scaled up as public funding alone is insufficient to meet the growing need for climate finance. International sources of private climate finance include multinational banks, global institutional investors like pension funds, insurers, asset managers, private equity and venture capital firms. These financiers co-invest in projects through blended finance mechanisms or public–private partnerships (PPPs). Initiatives like the Climate Finance Leadership Initiative (CFLI) [10] and the Glasgow Financial Alliance for Net Zero (GFANZ) mobilize cross-border investments and private capital for climate action [30]. Domestic sources for private climate finance include local banks, microfinance institutions, and commercial financial institutions. National green banks and financial intermediaries also channel private investment into climate-related projects through financial instruments such as green loans, sustainable bonds, and climate-focused investment funds. Individual households are also becoming important as they contribute funds to invest in residential projects such as rooftop solar, electric vehicles and energy-efficient appliances, which drive market transformation.

4.2.5 Philanthropic Capital

Philanthropic organizations and donations from foundations, charitable trusts, and high-net-worth individuals are becoming important sources of climate finance. Philanthropic capital often supports early-stage projects that can later be scaled by public and private investors. Individual contributions from family offices provide financial support for climate-focused NGOs is growing and are set to play an important role in advancing climate action. Philanthropic contributions from the Rockefeller Foundation, Bloomberg Philanthropies, the Bezos Earth Fund, and other donors help catalyze additional public and private investment. Individual philanthropists like Bill Gates support climate action through 'Breakthrough Energy' and MacKenzie Scott has channeled philanthropic capital through 'Yield Giving' to support climate-focused civil society organizations. While philanthropic finance represents a relatively small share of overall climate finance flows, it provides flexible capital to support projects that are often difficult to finance through traditional channels. It also delivers high impact as it fills critical financing gaps and mobilizes broader public and private climate finance.

4.3 Instruments for Climate Finance

The main instruments which are used to raise finance for climate-related activities are as follows.

4.3.1 Debt

Debt is the major instrument of climate finance and can be in the form of loans or bonds. Loans can be commercial or concessional. Commercial loans have a market-based interest rate and are delivered by commercial entities such as banks and financial institutions. These include project finance loans for specific commercial projects such as renewable energy parks or green hydrogen projects. Commercial loans also include individual consumer loans for household consumers and sovereign loans to State-Owned Entities (SOEs) for larger investments. Concessional loans have less stringent conditions than market loans. This could be at lower interest rates or longer repayment terms or a combination of these.

Bonds are other financial instruments that can be issued to raise climate finance. Bonds have a predefined maturity date, coupon rate, issue price, face value and other characteristics. Specifically, green bonds are fixed-income instruments whose proceeds are earmarked for projects with climate benefits, and the issuer must report on the impact of the projects. They have similar credit terms as conventional bonds, and the amount is paid back to the lender at a fixed time and interest rate which are slightly lower than conventional bonds. They can be issued by governments, municipalities and corporates to finance projects and activities that support climate change adaptation and mitigation.

4.3.2 Equity

Climate finance can be raised through equity in exchange for ownership of a company. Private investors, venture capitalists and corporate actors provide capital in return for an equity stake. Investors have an expectation of realizing market-level, risk-adjusted returns from dividends and sale of equity stakes in the company. Project-level equity is specifically for financing an earmarked project. A special purpose vehicle (SPV) is set up for the project to isolate the financial risk from the parent company. Company-level equity refers to instruments used to raise funds for an organization's transition. It involves securing capital to scale business models and technologies that are not specific to a project but fund the activities of the entire company and are reflected on its balance sheet.

4.3.3 Grants

Money given for climate change activities that does not need to be repaid is termed as grants. These are given when the commercial potential of activities is low or to prepare projects for investment readiness. Main sources of grants to address climate change are multilateral climate funds, MDBs, bilateral climate finance, national climate funds and philanthropic foundations. Apart from financial grants, technical assistance can also be provided for building institutional and stakeholder capacity, sharing expertise and consultations for a climate-related activity. Grants are also provided for project preparation activities to attract concessional finance and private capital.

4.3.4 Other Instruments

(a) Debt-for-Climate Swaps

This involves the sale of a foreign exchange debt to an investor or forgiveness of debt by the creditor, with a condition that the debt relief would have to be invested in climate change related activities. Through debt-for-climate swaps, debt payments are redirected in local currency to fund domestic climate projects, easing fiscal pressure on the country without increasing external debt. Bilateral debt swaps are direct arrangements between debtor countries and creditors, while tripartite swaps involve a third party such as an NGO or an SPV. In this arrangement, the NGO facilitates a loan to the debtor country at below-market interest rates to use the proceeds to repurchase debt held by commercial creditors. In return, the debt relief must be allocated to finance pre-agreed climate projects. As an example, Ecuador implemented a debt-for-climate swap in 2023 through the Galápagos marine bond, replacing over US$1.6 billion in international bonds with a US$656 million loan for marine conservation efforts [17].

(b) Guarantee

A guarantee transfers the risk of non-payment of a loan or similar debt instrument from the borrower or lender to a third party such as an MDB. Guarantees enable access to finance by leveraging the guarantor's higher credit rating to reduce borrowing costs or extend loan tenors. Governments may also act as sovereign guarantors for private capital. Guarantees differ in risk allocation and exposure coverage [18]. Partial credit guarantees cover part of the debt to cover commercial risks of non-payment. First loss guarantees cover early losses when blended finance is used. Loan portfolio guarantees involve clubbing the portfolio of loans and covering part of the losses on the full portfolio rather than one specific loan. The World Bank Group (WBG) offers about 20 different types of guarantees covering political risk, credit risk, and short-term trade finance, which can be used to catalyze climate finance [62].

(c) Viability Gap Funding

Viability gap funding (VGF) mechanism is usually a grant or a concessional loan to close the financing gap between the total cost of the project and its expected commercial returns. Without VGF, a project cannot be realized. Subordinated debt, which is a low-interest long term loan below the commercial rates, is one form of VGF. Another example is credit guarantees by MDBs to national financial institutions. This leads to a higher credit rating for project bonds, which allows the developer to avail lower interest rates from investors [9].

(d) Blended Finance

This approach uses catalytic capital from public or philanthropic sources to mobilize private capital for climate-related projects. It is used to attract commercial capital by derisking investments for climate projects while providing financial returns to investors. Blended finance uses public or concessional funds strategically to derisk projects that may not be implemented under normal market conditions through the use of concessional finance, subordinated loans, risk guarantees, grants and equity contributions with return expectations below the market rate of interest. Blended finance infrastructure deals attracted 40% of private capital for every US$1 worth of public or philanthropic money during 2021–2023, and around 10% of them mobilized more than US$2 for every US$1 invested, making blended finance highly impactful [92].

(e) Results-Based Climate Finance

Funds which are disbursed by a donor or investor after the pre-agreed results of the climate activity are achieved and verified are known as result-based climate finance (RBCF). This mechanism appeals to project donors as the payments are released only after third-party auditors confirm that the emission reduction from the project is real, permanent, and additional. RBCF helps in derisking projects and drawing in additional private sector investment. About 95% of international public climate finance is provided upfront before a project is operational. The Scaling Climate Action by Lowering Emissions (SCALE) initiative brings together all RBCF programs housed at the World Bank to incentivize low-carbon development [91].

(f) Insurance and Risk Management Mechanisms

These mechanisms absorb the residual physical and financial risks caused by climate change and protect against losses caused by extreme weather events and climate-related disasters. In a climate insurance product, investors earn returns from the ongoing policy premiums but face the risk of losing the invested amount if a disaster triggers a payout. 'Parametric insurance' involves payouts which are triggered when a parameter (e.g., rainfall, flood, droughts) crosses a predetermined limit rather than undertaking a detailed on-the-ground damage assessment. This lowers administrative costs such as the need for lengthy claims and ensures fast financial relief for agribusinesses, making it highly effective for rural households.

'Catastrophe Bonds' are high-yield debt instruments designed to transfer financial risks associated with natural disasters from insurers to capital market investors [23]. Institutional investors provide capital for insurance policies to protect against natural disasters and investors receive regular coupon payments. Insurance claims are paid out in case a specific climate event exceeds a pre-defined threshold. 'Sovereign Risk Pools' aggregate risks by combining exposure to projects with lower risk and volatility, reducing reinsurance costs. Regional risk pools like the *African Risk Capacity and Caribbean Catastrophe Risk Insurance Facility* allow governments to pool their risks and secure funds immediately after a natural disaster. Funds are maintained jointly for immediate first losses, and the higher risk layers are transferred to global markets [23].

4.4 Barriers for Climate Financing

4.4.1 Challenges in Mobilizing Climate Finance

Capitalizing climate finance is challenging due to several reasons. There is high competition for the availability of capital. Returns are generally lower from climate-focused projects, especially for climate adaptation, as compared to other high-return-yielding conventional investment options. Climate mitigation, adaptation, and resilience projects also have higher risk to capital, particularly in emerging and developing economies, resulting in higher interest rates for capital borrowings. Insufficient risk-mitigation tools and non-availability of suitable financing instruments also contribute to the climate investment gap, especially in emerging markets and developing economies.

4.4.2 Investment Risks

High political uncertainty leads to frequent policy changes. Lack of clarity in regulations creates delays and inconsistent climate policies undermine investor confidence. Exchange-rate volatility, rapid increases in inflation, and increasing sovereign debts result in higher cost of capital, especially in developing countries. High exposure to loans leads to a higher impact of volatile interest rates. The fast pace of change makes technology obsolete and creates risks of technology lock-in. Market risks include the risk of stranded assets for high-emission infrastructure due to changes in environmental regulations. Long supply chains add to investment risks as it increases the price volatility of green technologies.

4.4.3 Market Barriers

Market barriers increase the perceived risks and limit private investments. These include high upfront costs, lack of financial incentives, poor risk-return profiles, data asymmetries, counterparty risks and high transaction fees, which restrict climate finance flows. Lack of an investment-ready, bankable project pipeline is a major limitation. Reasons include small scale of the projects, unproven revenue streams, lack of creditworthiness and absence of counterparties. Underdeveloped local capital markets lead to reliance on foreign-denominated debt, increasing investment risks. Lack of concessional capital due to real and perceived risks prevents projects from achieving financial close as investors demand high risk-adjusted returns. Access to global capital markets also acts as market barriers that restrict the flow of funds from global institutional investors into emerging markets. Unlike mitigation projects which have a predictable revenue stream, adaptation projects which increase resilience, such as climate-resilient agriculture, do not produce direct revenue and without concessional finance these projects become unviable for commercial investors.

4.4.4 Structural and Capacity Barriers

Structural and capacity barriers include constraints arising from institutional, governance, technical, and human capacity that prevent effective mobilization and deployment of climate finance. Weak institutions, insufficient climate-related data, limited technical expertise, lack of project preparation facilities (PPF), poor governance and weak monitoring, reporting, and verification (MRV) systems lead to limited mobilization of climate finance. Inadequate project development capacity and limited technical expertise within local authorities, financial institutions, project developers and implementing agencies in low-income countries add to these barriers and discourage private investment by increasing the perceived risks and uncertainty. These structural barriers need to be addressed through long-term institutional reforms and sustained capacity development.

4.4.5 Policy Barriers

Policy barriers for climate finance are created by the poor design, inconsistency, or gaps in implementation of government policies. This reduces the attractiveness and effectiveness of climate-related investments. These include insufficient policies to attract private capital, weak policy signals, inconsistent carbon pricing and continuation of fossil fuel subsidies. Frequent changes in policies impact investments in projects which require a long gestation period and investors refrain from investing in these jurisdictions as returns are uncertain. Absence of sectoral targets, decarbonization roadmaps and credible transition pathways act as weak policy signals which lower investor confidence in future demand for low-carbon projects. Low carbon taxes or lack of carbon pricing weakens incentives to invest in low-carbon projects as climate costs are not internalized. Lack of green taxonomies, standards and climate disclosure requirements act as barriers to channel finance into green projects. Restrictions on capital movement and foreign exchange convertibility discourage private international climate finance flows. As an example, unclear electricity grid access rules, complicated tariff structures, weak renewable energy integration policies and unclear power market regulations increase the barriers for clean energy investments for renewable electricity. Poor coordination across government ministries and lack of transparency also undermines the credibility of the investment environment, resulting in lower climate finance flows.

Faster climate action requires higher private capital flows into climate-related projects. Public finance has a responsibility to advance development projects which build climate resilience and address adaptation concerns, but private investors prioritize higher risk-adjusted returns mostly through climate mitigation projects. The challenge also lies in channeling climate finance into credible, de-risked, investment-ready opportunities at scale, especially in emerging and developing economies.

4.5 Key Actions for Bridging the Climate Finance Gap

The sixth session of the Conference of the Parties serving as the meeting of the Parties to the Paris Agreement (CMA), which took place in Baku, Azerbaijan, from 11 to 24 November 2024, highlighted that current climate finance flows remain far below what is required, especially for adaptation. Developing countries' NDC implementation needs are estimated at US$5.1–6.8 trillion by 2030, and annual adaptation needs stood at US$215–387 billion per year [83]. It acknowledged the critical financing gap and committed to significantly scale up adaptation finance. The CMA decision also recognized the growing loss-and-damage needs and the requirement for enhanced financial support.

4.5.1 Aligning Global Financial Flows with the Paris Agreement Goals

As part of the NCQG at COP29, developed countries committed to take the lead in providing and mobilizing 300 billion USD for developing countries by 2035. Efforts need to be undertaken to fulfil this commitment at the earliest to mobilize the total international climate finance flows from private investors, multilateral banks and voluntary contributors to US$1.3 trillion. The CMA decision also agreed to triple the annual outflows of climate funds from 2022 levels by 2030 (approximately to US$5.2 billion annually), from operating entities of the UNFCCC Financial Mechanism and associated dedicated climate funds such as the GCF, GEF, Adaptation Fund, LDCF and SCCF to significantly scale the share of finance delivered through them.

4.5.2 Reforming the Financial System

The CMA decision emphasized the need to reform the global financial architecture to unlock higher climate finance flows. Efforts must be undertaken to lower the cost of capital, especially in developing countries, provide funds in the form of grants rather than high-interest loans, lower transaction costs and administrative barriers to access climate finance. Enhancing liquidity and market stability tools is also required to establish a global financial safety net that offers timely, unconditional financing support to all countries in need. Securing fiscal and policy space, including addressing unsustainable sovereign debt, is equally important. Rebalancing global economic governance with a more representative and accountable system that reflects all countries' interests and needs is also required [78].

Multilateral Development Banks (MDBs) reforms include updating governance and lending frameworks to expand concessional loans and grants, optimizing balance sheets and capital adequacy frameworks, expanding low-interest loan portfolios for emerging economies, higher use of guarantees and risk-sharing instruments and mobilizing private capital through blended finance platforms. Structural bottlenecks like overly rigid conditions and high transaction costs must be addressed through simplified application procedures, reduced reporting burdens and greater direct access to funds. MDBs are increasing their climate finance ambition through expanding their portfolio of concessional and grant-based finance and are encouraged to take higher risks to unlock private investment. Use of financing instruments such as first-loss guarantees, local currency financing and currency hedges is important to reduce risks and crowd in private capital. Pooling of programs and aggregation of projects is also useful instead of fragmented approaches.

4.5.3 Reducing the Cost of Capital in Emerging Markets

Several developing and LDC countries face financing costs that are two to four times higher than in advanced economies due to higher perceived investment risk. Lowering the cost of capital is therefore extremely important to accelerate investment in climate projects, especially in emerging and developing economies. Measures to mitigate currency risk include local currency debt, foreign exchange hedging mechanisms including currency swap arrangements and establishing MDB-backed currency risk products. For clean energy projects, good policy and regulation, reliable payments, timely permitting, availability of infrastructure, and tailored support for new and emerging technologies help to lower the cost of capital [52].

Other suggestions to overhaul the climate finance ecosystem include reforming credit rating methodologies, lengthening financing maturities to reduce rollover risk, establishing global and regional lenders of last resort to provide liquidity in times of stress, building local capital markets to reduce EMDE dependence on volatile foreign-currency borrowing, scaling blended finance tools to de-risk investments, and realigning institutional investor incentives to invest in high-growth, climate-aligned opportunities in EMDEs [72].

4.5.4 Enhancing Transparency and Accountability in Climate Finance

Transparency and accountability are essential to track climate finance and assess the progress on climate finance goals. All Parties to the UNFCCC now follow the Enhanced Transparency Framework (ETF) under the Paris Agreement as the mechanism to measure, report, and track climate action. Signatory countries have to submit Biennial Transparency Reports (BTRs) every two years, providing detailed data on climate finance received and provided. This will be the primary mechanism for tracking climate finance flows going forward. The BTRs are subject to a review by independent experts, which ensures consistency and accuracy of the reported numbers. The data collected through the ETF would form the basis of the Global Stocktake (GST), conducted every five years, for reviewing the progress and to identify future financial commitments. The UNFCCC's 'Standing Committee on Finance' will also produce biennial progress reports on collective global progress beginning in 2028, and a dedicated assessment of access to climate finance will take place in 2030.

Development of standardized definitions and taxonomies for climate finance with a clear distinction between sources and destinations of finance (for example, adaptation and mitigation finance, public and mobilized private finance) would be useful for reporting consistency. Globally agreed definitions would enable adoption of harmonized reporting frameworks where common methodologies are applied using internationally agreed standards across MDBs, bilateral and multilateral donors, public and private funders.

Currently, climate finance is tracked through a decentralized network of publicly maintained dashboards and privately held databases. The Climate Policy Initiative maintains a 'Climate Finance Data Dashboard' which provides a comprehensive overview of global climate finance flows [20]. For public funds, the UNFCCC 'Climate Finance Data Portal' provides an open-data source which serves as a central gateway to track financial resources to monitor the mobilization of global climate finance for developing countries [88].

To address the lack of a single centralized global digital climate finance registry, UN Climate Change in partnership with Microsoft and EY, has developed the UNFCCC 'Climate Data Hub'. It provides a centralized source of global climate data to support climate action and enhances data transparency. The interactive web portal provides easy access to data related to financial, technology development and transfer and capacity-building (FTC) support, as reported under the ETF. It provides datasets to track financial support provided through bilateral, regional and other channels, financial support provided through multilateral channels, financial support mobilized through public interventions, financial support needed, financial support received, Article 13 and transparency-related support needed and support received [89].

Nevertheless, systemic challenges on data fragmentation and inconsistency; capacity disparities related to data-collection methodologies, MRV and institutional infrastructure among nations; and siloed historical reporting framework and data workflows continue to exist. Efforts are also required to transition from biennial or irregular periodic updates to a more frequent reporting system to track climate finance flows and build granular visualization tools for tracking sectoral climate finance flows.

Initiatives like the 'ICAT Climate Finance Transparency Guide', which provides methodological guidance to establish and implement national climate finance transparency frameworks for developing countries, are a promising step in this direction. The guide offers a step-by-step approach to track, measure, manage and report on climate finance flows, aligning with the reporting requirements under the ETF [49].

Improving transparency of private finance flows is also important. This requires project-level disclosures including information on beneficiaries, financing instruments, updates on implementation status, expected climate impacts and financial integrity checks. Internationally agreed methodologies to measure private capital mobilization and linking them to measurable outcomes such as emissions reduction and impact on resilience are also required.

4.5.5 Develop Bankable Project Pipelines

The climate finance gap is actually a bankability gap due to the lack of investment-ready bankable projects. This gap can be bridged through targeted interventions. Supporting project preparation through funding feasibility studies, technical assistance, financial structuring, assessment of regulatory compliance and preparation of engineering design, reduces early-stage risks. Dedicated PPF from MDBs and development organizations helps to establish the financial viability of the project through revenue modeling, offtake agreements and cost structuring. Other actions include creating climate project databases, strengthening project preparation capabilities, standardized contracts and reporting frameworks, improved climate taxonomies and developing domestic financial sector capacity.

One example is the GCF's *Readiness and Preparatory Support Programme,* which supports country-driven initiatives to strengthen their institutional capacities, governance mechanisms, and planning frameworks for advancing the climate action agenda [42]. Early assistance in establishing technological reliability, conducting Environmental Impact Assessment (EIA) and Social Impact Assessment (SIA), obtaining permitting approvals and developing a contractual framework helps to derisk and transform a climate project into a bankable investment.

Bundling of multiple projects through aggregation mechanisms, use of SPVs and well-designed securitization structures transforms smaller projects into investable portfolios and attracts institutional investors. Creation of a national climate investment platform streamlines development of project pipelines at the national level. This helps in standardizing project screening, prioritization, and reporting frameworks and creates a pipeline of future investment opportunities. Vetting of these projects by a centralized governance body comprising key ministries and government departments provides government backing and lowers political and policy risks. The platform also becomes a dynamic database of commercially viable projects which can be used by project developers and financial institutions for investment.

Several new country and regional platforms for climate and nature finance were announced at COP30, which would be supported by the Green Climate Fund's *Readiness Programme* [36]. *Brazil Country Platform* and *Caribbean Regional Platform* are already functional as country platforms for climate finance. They support PPF to develop a strong project pipeline and provide assistance for mobilizing funding to implement climate projects [43]. Another example is the Climate Investment Platform (CIP), which is a joint initiative of the International Renewable Energy Agency (IRENA), United Nations Development Programme (UNDP), and Sustainable Energy for All (SEforALL), in collaboration with the GCF. The platform aims to increase capital mobilization for the development of renewable energy projects through tailored technical assistance and facilitation support to reach commercial feasibility readiness and provide project matchmaking support with financial institutions.

Creation of national green banks and specialized country-driven financial institutions that use public funds to help generate catalytic financing by absorbing risk and mobilizing private investment for climate projects. Examples include the *Clean Energy Finance Corporation* (Australia), which is a standalone green bank; the *Indian Renewable Energy Development Agency* (India), which is a government-backed non-banking financial institution; the *UK Infrastructure Bank* (United Kingdom) and the *Malaysian Green Technology and Climate Change Centre* (Malaysia), which is a hybrid national facility to support financing of projects.

4.6 Successful Models and Examples for Mobilization of Climate Finance

4.6.1 CIF Capital Markets Mechanism

In early 2025, the Climate Investment Funds (CIF) pioneered the model for raising investment capital directly from international capital markets. The CIF Capital Markets Mechanism (CCMM) raised US$500 million from global investors through the issuance of a 3-year bond, which was listed on the International Securities Market at the London Stock Exchange. It mobilized private capital from global investors, including asset managers, insurance, pension funds, which contributed 51%, central banks and official institutions (36%) and commercial banks and corporates (13%). The coupon rate of 4.75% would be payable semi-annually and the proceeds from the sale of the bond will be used to fund projects in developing countries through the CTF [13]. The World Bank acts as the treasury manager and trustee of the CCMM, overseeing bond issuances, liquidity management, financial risk management, credit rating management, and financial reporting [14]. The sale of fixed-income securities provides institutional capital for early-stage investments for climate projects without waiting for long-term reflows of investment from these projects. This funding is channeled into projects on different continents through six MDBs which act as implementing entities of the CTF.

4.6.2 Innovative Finance Facility for Climate in Asia and the Pacific

Innovative Finance Facility for Climate in Asia and the Pacific (IF-CAP) is a multi-donor financing partnership facility and the first leveraged guarantee mechanism for climate finance to be adopted by an MDB. It aims to strengthen member countries' capacity to build a larger pipeline of potential investments. The guarantees cover payment defaults in a reference portfolio which is sized at US$10 billion, diversified across 29 member countries.

Donor partners such as Denmark, Japan, the Republic of Korea, Sweden, the United Kingdom and the United States, and other public, private and philanthropic financing partners provide capital for a trust fund which acts as a guarantee facility to backstop repayments, covering payment defaults for a subset of ADB's loans, thereby taking the risk off the ADB's balance sheet. This frees up capital to increase lending for climate investments. Supplementary grants are provided to facilitate project preparation, capacity building, and knowledge solutions. It is expected that for every US$1 that is guaranteed, up to US$4.5 of new climate finance will be generated through this mechanism. Over an initial 5-year investment period, the ADB hopes to raise US$3 billion in guarantees from financing partners, which could enable as much as US$15 billion for climate projects across the region [8].

The success of the IF-CAP model demonstrates that bilateral donors can utilize limited contributions from development cooperation budgets to unlock financing within an MDB. The guarantees provided can leverage the existing balance sheets and can be replicated by other regional development for unlocking the available capital for climate projects [66].

4.6.3 Just Energy Transition Partnerships

Just Energy Transition Partnerships (JETPs) are multilateral platforms between developed and emerging economies for delivering climate finance for accelerating the transition to an equitable and socially inclusive low-carbon economy [45]. The JETPs emphasize justice and equity for the energy transition and aim to expedite coal phase-out by replacing it with renewable energy. JETPs were established between G7 + countries and South Africa (2021), Indonesia (2022), Vietnam (2022), and

Senegal (2023). The International Partners Group (IPG), composed of G7 countries and international financial institutions, pledged approximately US$46.5 billion to support these JETPs, but the United States withdrew its support in February 2025 [76]. Funding is provided through a mix of public and private grants, concessional loans, commercial debt, and equity investments, but only 3–4% of financing is in the form of grants, which is concerning as it could lead to higher debt burdens for countries [59].

A key highlight of JETP is the benefit and/or revenue sharing models for impacted communities. JETP Indonesia has operationalized its 'Just Transition Framework', which emphasizes economic diversification and transformation as the approach to mitigate risks for stakeholders and communities [74]. The model is built on the four pillars of a just energy transition, which are recognitional justice, restorative justice, distributive justice, and procedural justice and ensures that the energy transition creates new economic opportunities while moving away from coal-dependent growth. It also addresses the risks of early retirement of coal workers through reskilling programs for transitioning to green jobs [19].

While flows of funds to JETPs have been slow, the initiative has high-potential for mobilizing public and private financing for early coal phaseout in EMDEs. It also provides a unique financing model for moving away from a coal-based economy to a cleaner energy mix in emerging economies [21].

4.6.4 *Green Sukuk*

Green Sukuk is a Shariah-compliant Islamic bond designed to finance environmentally sustainable projects. Investors earn returns through profit-sharing or income streams tied to an underlying tangible asset [48]. Like a conventional green bond, the proceeds are exclusively channeled toward projects with positive environmental benefits such as renewable energy, clean transportation, green buildings and sustainable infrastructure. Green Sukuk mobilized about US$11 billion by the third quarter of 2024, and it is estimated that US$30–50 billion could be raised through the financial mechanism [79]. In October 2025, the Islamic Development Bank (IsDB) raised €500 million through its Green Sukuk issuance under its enhanced 2025 Sustainable Finance Framework. This followed two other successful public US$ benchmark transactions that raised a total of US$3.5 billion from the global capital markets earlier in 2025 [55].

Countries such as Malaysia, Indonesia, Saudi Arabia and the UAE have developed regulatory frameworks and taxonomies to facilitate the issue of Green Sukuk and have introduced tax incentives and grant schemes. Green Sukuk have the potential to raise climate finance, especially within Muslim-majority regions, and bridge the climate finance gap through ethical financing aligned with Shariah principles. For investors, it offers exposure to new markets and an opportunity to expand the investor base for climate finance. Challenges include legal enforceability, governance issues, greenwashing risks and technical issues for enhanced impact measurement using digital tools [90].

4.6.5 *Orange Capital Instruments*

The 'Orange Movement' is a global initiative mobilizing capital to empower women, girls, and gender minorities as leaders in climate action, sustainable peace, and economic prosperity. Orange bonds, loans, grants and funds are the world's first asset class for gender-lens investing [68]. Similar to 'Green Bond Principles', which built momentum for climate investments, 'Orange Bonds Principles' create a robust framework to guide capital towards gender-lens investing at scale. Impact Investment Exchange (IIX) issued the first Women's Livelihood Bonds (WLBs) in 2017 and raised US$100 million in 2023 through this bond [27]. The Orange Bond Initiative (OBI) seeks to mobilize US$10 billion to empower 100 million women, girls, and gender minorities by 2030, and has gained momentum with 76,000 signatories from six continents [73]. In September 2025, ITOCHU Corporation issued Orange Bonds worth US$102.5 million to invest in companies proactively supporting and empowering women in Japan and Guatemala [56]. Orange bonds and similar capital instruments provide a specific classification focused on gender equality and address the multifaceted challenges at the intersection of gender with race, climate, and economy. These instruments advance gender-just prosperity, peace, climate resilience and enhance inclusion in decision-making and are a useful addition to gender- and climate-focused financial products.

4.7 Conclusion

Climate finance is a key enabler for climate action but is significantly below what is required to achieve the goals of the Paris Agreement. The global climate finance gap is not due to lack of available capital, but reflects a collective failure to mobilize and direct capital for addressing climate change. Emerging economies and LDCs require greater access to affordable climate finance to address climate mitigation, adaptation, and resilience. However, the existing climate finance ecosystem is fragmented, uncoordinated and dominated by irregular investments which are heavily dependent on donors and public financing.

The scale and the urgency of the climate challenge require collective action from all stakeholders, including multilateral development banks, climate funds, philanthropic organizations, and national governments. Private capital mobilization is essential to bridge the climate finance gap and public finance must be leveraged through innovative financial instruments to mobilize capital and unlock investment at scale. Reducing the cost of capital, addressing risks, strengthening institutions, capacity-building and creating enabling policy environments are essential for mobilizing climate finance. Successful global examples and models demonstrate that barriers can be overcome to raise and deploy climate finance effectively for climate-related projects. Bridging the climate finance gap requires urgent and sustained action today to ensure that climate finance becomes the catalyst that accelerates the global transition to a net-zero and climate-resilient future.

References

1. Adaptation Fund (2026) Adaptation Fund: Year-in-Review 2025. https://www.adaptation-fund.org/adaptation-fund-year-in-review-2025/. Accessed 25 June 2026
2. Adaptation Fund (n.d.) Direct Access. https://www.adaptation-fund.org/about/direct-access/. Accessed 08 April 2026
3. Adaptation Fund, Climate Investment Funds, Global Environment Facility and Green Climate Fund (2025) Multilateral Climate Funds Joint Results Report. https://www.adaptation-fund.org/multilateral-climate-funds-joint-results-report/. Accessed 10 April 2026
4. African Development Bank (2025) The African Development Bank Group's Climate Action and Financing. COP30 Focus Africa. https://www.afdb.org/en/cop30/focus-africa/african-development-bank-groups-climate-action-and-financing. Accessed 08 April 2026
5. Agence Française de Développement (n.d.) Climate. https://www.afd.fr/en/themes/climate. Accessed 11 April 2026
6. Asian Development Bank (2026) Climate Finance in 2025. Infographic. https://www.adb.org/news/infographics/climate-finance-2025. Accessed 25 June 2026
7. Asian Development Bank (n.d.) ADB's Work on Climate Change and Disaster Risk Management. https://www.adb.org/what-we-do/topics/climate-change/overview. Accessed 08 April 2026
8. Center for Global Development (2023) IF-CAP Recap: The Asian Development Bank's Big Climate Finance Bet. https://www.cgdev.org/blog/if-cap-recap-asian-development-banks-big-climate-finance-bet. Accessed 25 April 2026
9. Cities Climate Finance Leadership Alliance (n.d.) Viability Gap Funding. https://citiesclimatefinance.org/financial-instruments/instruments/viability_gap_funding. Accessed 20 April 2026
10. Climate Finance Leadership Initiative (n.d.) About. https://www.bloomberg.com/cfli/about/. Accessed 15 April 2026
11. Climate Funds Update (n.d.) Global Climate Change Alliance. https://climatefundsupdate.org/the-funds/global-climate-change-alliance/. Accessed 15 April 2026
12. Climate Funds Update (n.d.) Indonesia Climate Change Trust Fund. https://climatefundsupdate.org/the-funds/indonesia-climate-change-trust-fund/. Accessed 12 April 2026
13. Climate Investment Funds (2025) CCMM Issuance. https://www.cif.org/news/ccmm-issuance. Accessed 25 April 2026
14. Climate Investment Funds (n.d.) CIF Capital Markets Mechanism (CCMM). https://www.cif.org/CCMM. Accessed 25 April 2026
15. Climate Investment Funds (n.d.) CIF Funding. https://www.cif.org/cif-funding. Accessed 25 June 2026
16. Climate Policy Initiative (2025) Global Landscape of Climate Finance 2025. https://www.climatepolicyinitiative.org/publication/global-landscape-of-climate-finance-2025/. Accessed 05 April 2026
17. Climate Policy Initiative (2026) Debt for Climate Swaps. https://www.climatepolicyinitiative.org/wp-content/uploads/2026/01/Debt-for-Climate-Swaps.pdf. Accessed 20 April 2026
18. Climate Policy Initiative (2026) Guarantees. https://www.climatepolicyinitiative.org/wp-content/uploads/2026/01/Guarantees.pdf. Accessed 25 June 2026

19. Climate Policy Initiative (2026) CPI Launches Report on Indonesia's Just Energy Transition Finance Framework. https://www.climatepolicyinitiative.org/press-release/cpi-launches-report-on-indonesias-just-energy-transition-finance-framework/. Accessed: 25 June 2026
20. Climate Policy Initiative (n.d.) Global Landscape of Climate Finance Data Dashboard. https://www.climatepolicyinitiative.org/resources/data-visualizations/global-landscape-of-climate-finance-data-dashboard/. Accessed 25 June 2026
21. Climate Policy Initiative (n.d.) Just Energy Transition Partnerships (JET-Ps). Climate Finance Reform Compass. https://compass.climatepolicyinitiative.org/themes/country-led-initiatives/jet-ps. Accessed 25 June 2026
22. Climate Policy Initiative (n.d.) Climate Finance Instruments Toolkit. https://www.climatepolicyinitiative.org/resources/data-visualizations/toolkit-instruments/. Accessed 20 April 2026
23. Climate Policy Initiative (n.d.) Top-Down Climate Finance Needs. https://www.climatepolicyinitiative.org/publication/top-down-climate-finance-needs/. Accessed 06 April 2026
24. Congressional Research Service (2025) U.S. Agency for International Development: An Overview. CRS Report IF10261, updated 5 September. https://www.congress.gov/crs-product/IF10261. Accessed 11 April 2026
25. Deutsche Gesellschaft für Internationale Zusammenarbeit (n.d.) Klimawandel. https://www.giz.de/de/expertise/klima-umwelt/klimawandel. Accessed 10 April 2026
26. Development Bank of Southern Africa (n.d.) Green Fund. https://www.dbsa.org/solutions/climate-financing/green-fund. Accessed 15 April 2026
27. Environmental Finance (2025) IIX Raises $100m from 'Largest' Gender Bond. https://www.environmental-finance.com/content/news/iix-raises-$100m-from-largest-gender-bond.html. Accessed 29 April 2026
28. European Bank for Reconstruction and Development (2026) EBRD Invests Record €16.8 Billion in 2025. https://www.ebrd.com/home/news-and-events/news/2026/ebrd-invests-record--16-8-billion-in-2025.html. Accessed 10 April 2026
29. French Facility for Global Environment (n.d.) About the French Facility for Global Environment. https://www.ffem.fr/en/themes/about-french-facility-global-environment. Accessed 11 April 2026
30. Glasgow Financial Alliance for Net Zero (n.d.) About. https://www.gfanzero.com/about/. Accessed 15 April 2026
31. Global Environment Facility (2025) GEF Approves $372 Million in Funding to Tackle Environmental Challenges Worldwide and Gears Up to New Cycle of Investments. https://www.thegef.org/newsroom/press-releases/gef-approves-372-million-funding-tackle-environmental-challenges-worldwide. Accessed 25 June 2026
32. Global Environment Facility (n.d.) GEF-9 Replenishment. https://www.thegef.org/who-we-are/funding/gef-9-replenishment. Accessed 25 June 2026
33. Government of Canada (2026) Evaluation of the International Climate Finance Program: Report. https://www.canada.ca/en/environment-climate-change/corporate/transparency/priorities-management/evaluations/evaluation-international-climate-finance-program.html. Accessed 11 April 2026
34. Government of Canada (n.d.) Access to Climate Finance. https://www.canada.ca/en/services/environment/weather/climatechange/canada-international-action/climate-finance/access.html. Accessed 11 April 2026
35. Government of Canada (n.d.) International Financing. https://www.canada.ca/en/environment-climate-change/services/climate-change/international-climate-financing.html. Accessed 11 April 2026
36. Green Climate Fund (2025) Fourteen New Country Platforms Announced in Collaboration with GCF to Drive Country-Led Climate Solutions. https://www.greenclimate.fund/about/news-and-updates/fourteen-new-country-platforms-announced-collaboration-gcf-drive-country-led-climate-solutions. Accessed 20 April 2026
37. Green Climate Fund (n.d.) About. https://www.greenclimate.fund/about. Accessed 08 April 2026

38. Green Climate Fund (n.d.) Resource Mobilisation: GCF-2. https://www.greenclimate.fund/about/resource-mobilisation/gcf-2. Accessed 25 June 2026
39. Green Climate Fund (n.d.) GCF's Second Replenishment (GCF-2). https://www.greenclimate.fund/gcf-2. Accessed 25 June 2026
40. Green Climate Fund (n.d.) Japan International Cooperation Agency (JICA). https://www.greenclimate.fund/ae/jica. Accessed 11 April 2026
41. Green Climate Fund (n.d.) Development Bank of Southern Africa (DBSA). https://www.greenclimate.fund/ae/dbsa. Accessed 15 April 2026
42. Green Climate Fund (n.d.) Readiness and Preparatory Support. https://www.greenclimate.fund/access-funding/readiness-preparatory-support. Accessed 20 April 2026
43. Green Climate Fund (n.d.) Country Platforms for Climate Finance. https://www.greenclimate.fund/document/country-platforms-climate-finance. Accessed 20 April 2026
44. HM Government (n.d.) International Climate Finance. GOV.UK. https://www.gov.uk/government/collections/international-climate-finance. Accessed 11 April 2026
45. Institute for Human Rights and Business (n.d.) What Are JETPs? (Just Energy Transition Partnerships). https://www.ihrb.org/resources/what-are-jetps-just-energy-transition-partnerships. Accessed 28 April 2026
46. Inter-American Development Bank (2024) IDB to Scale Annual Climate Financing to $11.3 Billion by 2030, Accelerating Impact. https://www.iadb.org/en/news/idb-scale-annual-climate-financing-113-billion-2030-accelerating-impact. Accessed 10 April 2026
47. Inter-American Development Bank (2026) Inter-American Development Bank Annual Report 2025: The Year in Review. https://publications.iadb.org/publications/english/document/idb_annual_report_2025_web_version.pdf. Accessed 10 April 2026
48. International Capital Market Association, Islamic Development Bank and London Stock Exchange Group (2024) ICMA-IsDB-LSEG Guidance on Green, Social and Sustainability Sukuk. https://www.icmagroup.org/assets/documents/Sustainable-finance/ICMA-IsDB-LSEG-Guidance-on-Green-Social-and-Sustainability-Sukuk-April-2024.pdf. Accessed 28 April 2026
49. International Climate Action Transparency (n.d.) Climate Finance Transparency Guide. https://climateactiontransparency.org/climate-finance-transparency-guide/. Accessed: 20 April 2026
50. International Climate Initiative (n.d.) The International Climate Initiative (IKI). https://www.international-climate-initiative.com/en/. Accessed 10 April 2026
51. International Climate Initiative (n.d.) Was ist die Internationale Klimaschutzinitiative (IKI)? [Video]. https://www.international-climate-initiative.com/en/iki-media/video/was-ist-die-internationale-klimaschutzinitiative-iki-new69720917e12d6070938874/. Accessed 10 April 2026
52. International Energy Agency (2024) Reducing the Cost of Capital: Executive Summary. https://www.iea.org/reports/reducing-the-cost-of-capital/executive-summary. Accessed 20 April 2026
53. International Monetary Fund (n.d.) Fossil Fuel Subsidies. https://www.imf.org/en/Topics/climate-change/energy-subsidies. Accessed 05 April 2026
54. International Monetary Fund (n.d.) World Economic Outlook: GDP, Current Prices. IMF DataMapper. https://www.imf.org/external/datamapper/NGDPD@WEO/OEMDC/ADVEC/WEOWORLD. Accessed 06 April 2026
55. Islamic Development Bank (n.d.) IsDB Issues Another Successful Green Sukuk Under Enhanced Sustainable Finance Framework. https://www.isdb.org/news/isdb-issues-another-successful-green-sukuk-under-enhanced-sustainable-finance-framework. Accessed 28 April 2026
56. ITOCHU Corporation (2025) ITOCHU Announces the Issuance of an Orange Bond with Funds Use Limited to Promote Women's Advancement. https://www.itochu.co.jp/en/news/press/2025/250905_2.html. Accessed 29 April 2026
57. Japan Bank for International Cooperation (no date) Role and Function. https://www.jbic.go.jp/en/about/role-function.html. Accessed 11 April 2026

58. Kamruzzaman M (2015) Introduction to the Bangladesh Climate Change Trust Fund, ICCCAD/Dhaka Tribune Articles. https://icccad.net/dhaka-tribune-articles/introduction-to-the-bangladesh-climate-change-trust-fund-2/. Accessed 15 April 2026
59. Karg A, Gupta J, Chen Y (2025) Just Energy transition partnerships: an inclusive climate finance approach? Energy Res Soc Sci 125, article 104103. https://www.sciencedirect.com/science/article/pii/S2214629625001847. Accessed 28 April 2026
60. KfW Entwicklungsbank (n.d.) KfW Development Bank. https://www.kfw-entwicklungsbank.de/International-financing/KfW-Entwicklungsbank/. Accessed 10 April 2026
61. Lok Sabha (n.d.) Unstarred Question No. 2465 [Reply]. https://sansad.in/getFile/loksabhaquestions/annex/186/AU2465_AE1h6g.pdf. Accessed 15 April 2026
62. Multilateral Investment Guarantee Agency (n.d.) World Bank Group Guarantee Platform. https://www.miga.org/world-bank-group-guarantee-platform. Accessed 20 April 2026
63. Naran B, Buchner B, Price M, Stout S, Taylor M, Zabeida D (2024) Global Landscape of Climate Finance 2024: Insights for COP29. Climate Policy Initiative (CPI). https://www.climatepolicyinitiative.org/wp-content/uploads/2024/10/Global-Landscape-of-Climate-Finance-2024.pdf. Accessed 29 April 2026
64. Nordic Development Fund (n.d.) Home. https://www.ndf.int/. Accessed 12 April 2026
65. Norway's International Climate and Forest Initiative (n.d.) About Us. https://www.nicfi.no/about-us/. Accessed 12 April 2026
66. OECD (2025) Innovative Finance Facility for Climate in Asia and the Pacific. Blended Finance Case Studies. https://www.oecd.org/en/publications/blended-finance-case-studies_2fb90b9a-en/innovative-finance-facility-for-climate-in-asia-and-the-pacific_4a7545b5-en.html. Accessed 28 April 2026
67. OECD (2026) Climate Finance Provided and Mobilised by Developed Countries in 2013–2024. Climate Finance and the USD 100 Billion Goal. OECD Publishing, Paris. https://www.oecd.org/en/publications/climate-finance-provided-and-mobilised-by-developed-countries-in-2013-2024_ab5eb9ad-en.html. Accessed 25 June 2026
68. Orange Movement (n.d.) Orange Capital. https://orangemovement.global/orange-capital. Accessed 29 April 2026
69. Philippines Climate Change Commission (n.d.) Climate Finance: People's Survival Fund. https://climate.gov.ph/our-programs/climate-finance/peoples-survival-fund. Accessed 15 April 2026
70. Philippines Department of Finance (n.d.) People's Survival Fund. https://psf.dof.gov.ph/. Accessed 15 April 2026
71. Rwanda Green Fund (n.d.) About FONERWA. https://www.greenfund.rw/backendpd/about. Accessed 15 April 2026
72. Sachs J, Sachs L, Camelo Vega A, Willis BM (2025) Lowering the Cost of Capital for Climate and SDG Finance in Emerging Markets and Developing Economies (EMDEs). Columbia Center on Sustainable Investment, New York. https://ccsi.columbia.edu/content/lowering-cost-of-capital-climate-sdgs-emdes. Accessed 25 June 2026
73. Shahnaz D (2025) Why the Future of Financial Inclusion Could Be Orange, I by IMD. https://www.imd.org/ibyimd/finance/why-the-future-of-financial-inclusion-could-be-orange/. Accessed 29 April 2026
74. Southeast Asia Energy Transition Partnership (n.d.) Operationalization of the Just Transition Framework in JETP Indonesia. https://www.energytransitionpartnership.org/projects/operationalization-of-the-just-transition-framework-in-jetp-indonesia/. Accessed 28 April 2026
75. Swiss Agency for Development and Cooperation (n.d.) Swiss International Climate Finance. https://www.sdc-cde.ch/en/swiss-international-climate-finance. Accessed 12 April 2026
76. Transparency International (n.d.) Strengthening Just Energy Transition Partnerships (JETPs): Lessons Learned for a Just Energy Transition. https://www.transparency.org/en/publications/strengthening-just-energy-transition-partnerships-jetps-lessons-learned-for-a-just-energy-transition. Accessed 28 April 2026
77. U.S. International Development Finance Corporation (n.d.) About Us. https://www.dfc.gov/who-we-are/about-us. Accessed 11 April 2026

78. UNCTAD (2025) All Roads Lead to Reform: A Financial System Fit to Mobilize $1.3 Trillion for Climate Finance. Geneva: UN Trade and Development. https://unctad.org/publication/all-roads-lead-reform-financial-system-fit-mobilize-13-trillion-climate-finance. Accessed 20 April 2026
79. UNDP (2025) Green Sukuk: A Tool for Sustainable Financing. https://www.undp.org/sites/g/files/zskgke326/files/2025-05/undp-kfh-green-sukuk-tool-for-sustainable-financing.pdf. Accessed 28 April 2026
80. UNEP (2025) Adaptation Gap Report 2025: Running on Empty. Nairobi: UNEP. https://www.unep.org/resources/adaptation-gap-report-2025. Accessed 06 April 2026
81. UNFCCC (2024) COP29 UN Climate Conference Agrees to Triple Finance to Developing Countries, Protecting Lives and Livelihoods. https://unfccc.int/news/cop29-un-climate-conference-agrees-to-triple-finance-to-developing-countries-protecting-lives-and. Accessed 06 April 2026
82. UNFCCC (2025) Report on the Baku to Belém Roadmap to 1.3T. https://unfccc.int/sites/default/files/resource/Relatorio_Roadmap_COP29_COP30_EN_final.pdf. Accessed 06 April 2026
83. UNFCCC (2025) New Collective Quantified Goal on Climate Finance. Decision-/CMA.6. https://unfccc.int/sites/default/files/resource/CMA_11%28a%29_NCQG.pdf. Accessed 20 April 2026
84. UNFCCC (n.d.) Introduction to Climate Finance. https://unfccc.int/topics/introduction-to-climate-finance. Accessed 05 April 2026
85. UNFCCC (n.d.) The Special Climate Change Fund (SCCF). https://unfccc.int/topics/climate-finance/resources/reports-of-the-special-climate-change-fund. Accessed 08 April 2026
86. UNFCCC (n.d.) Least Developed Countries (LDC) Fund. https://unfccc.int/process-and-meetings/bodies/funds-and-financial-entities/least-developed-countries-ldc-fund. Accessed 08 April 2026
87. UNFCCC (n.d.) Rwanda Green Fund–FONERWA. https://unfccc.int/climate-action/momentum-for-change/financing-for-climate-friendly-investment/rwanda-green-fund-fonerwa. Accessed 15 April 2026
88. UNFCCC (n.d.) Climate Finance Data Portal. https://unfccc.int/climatefinance?home. Accessed: 25 June 2026
89. UNFCCC (n.d.) Financial, Technology and Capacity-Building (FTC) Support Dataset. https://data.unfccc.int/datasets/ftc. Accessed 20 April 2026
90. Wan Zahari WMZ (2025) Operationalising Green Sukuk within the world bank environmental and social framework. Discov Environ 3, article 288. https://doi.org/10.1007/s44274-025-00490-6. Accessed 29 April 2026
91. World Bank (2022) What You Need to Know About Results-Based Climate Finance. https://www.worldbank.org/en/news/feature/2022/08/17/what-you-need-to-know-about-results-based-climate-finance. Accessed 20 April 2026
92. World Bank (n.d.) Blended Finance. Public-Private Partnership Legal Resource Center. https://ppp.worldbank.org/blended-finance. Accessed 20 April 2026
93. World Bank Group (2025) World Bank FY25 Climate-Related Disclosures. https://thedocs.worldbank.org/en/doc/f8a2c887929c44d9dab187f3fc1ad293-0040012025/original/WB-FY25-Climate-Related-Disclosures.pdf. Accessed 08 April 2026
94. World Economic Forum (2023) Climate Change is Costing the World $16 Million per Hour. https://www.weforum.org/stories/2023/10/climate-loss-and-damage-cost-16-million-per-hour/. Accessed 08 April 2026

Chapter 5
Delivering Climate Solutions Through Markets

Abstract This chapter describes the role of markets in delivering climate solutions. It highlights the important role of carbon markets and describes the compliance markets, voluntary carbon markets and Article 6 mechanisms for carbon trading. It presents how markets can be used for incentivizing renewable energy generation, for enhancing resource efficiency and environmental protection using global examples. Market solutions based on natural capital and examples of nature-based markets are presented. Market mechanisms that operationalize carbon dioxide removal credits and chain of custody models are discussed using functional examples from different subsectors.

Keywords Market mechanisms · Carbon trading · Renewable energy certificates · Carbon dioxide removal credits · Chain of Custody

5.1 Introduction

Market-based mechanisms are economic instruments that use incentives and market signals to trigger environmentally beneficial actions. Unlike regulatory instruments, they create a financial motivation to act by penalizing GHG emissions or other environmental externalities. Markets bring in economic efficiency and deliver cost-effective emission reduction as compared to traditional command-and-control measures. It puts a price on emissions and shifts the costs of externalities onto the polluters. Market mechanisms provide flexibility in achieving environmental goals and are technology-neutral. Interventions are undertaken where they are cheapest, minimizing overall compliance costs. Markets also incentivize technological innovation and its deployment, which leads to cost reduction in a positive reinforcing loop. Market mechanisms provide an additional revenue stream which helps to mobilize private capital, which is essential for scaling climate action.

This chapter describes the role of markets to accelerate climate action and deliver climate solutions. Section 5.1 commences with identifying the three different types of

K. Narula, *The Climate Catalyst*, Advances in Sustainability Science and Technology,
https://doi.org/10.1007/978-981-92-5213-8_5

market mechanisms—quantity-based approach, price-based approach and baseline-and-crediting mechanisms and presents a short evolution of market mechanisms. Section 5.2 focuses on carbon markets and describes the compliance markets, voluntary carbon markets and Article 6 mechanisms for carbon trading. Section 5.3 presents different forms of energy attribute certificates and their characteristics using global examples. Section 5.4 describes how markets can be used for enhancing resource efficiency and environmental protection using the examples of energy efficiency certificates, circular economy credits and circular credit mechanism. Market mechanisms based on natural capital are presented in Sect. 5.5, followed by examples of nature-based and farming markets in Sect. 5.6. Section 5.7 focuses on carbon dioxide removal credits and describes direct air capture, biochar, enhanced rock weathering, bioenergy with carbon capture and storage, and marine carbon dioxide removal credits. Section 5.8 explains different chain of custody models such as mass-balance and book and claim system giving functional examples and how these approaches can facilitate adoption of climate solutions. The last section summarizes how markets can accelerate climate action.

5.1.1 Types of Market Mechanisms

Market mechanisms can be divided into three different types. In a quantity-based approach, the government sets the quantity of permissible emissions and the market determines the price. If applied rigorously, the environmental outcome, such as GHG emission reduction, is certain, but the price depends on market dynamics of supply and demand. This includes cap and trade systems such as the European Union (EU) Emission Trading Scheme (ETS) and voluntary carbon offset markets where carbon credits are generated for emission reduction below the established baseline.

In a price-based approach, the government sets the price of a unit of emission and the market determines the quantity of reduced emissions. In this approach, the price is fixed, but the total quantity of emission reductions is uncertain. Price-based approaches include carbon taxes, which set a carbon price exogenously to drive domestic decarbonization. Carbon border adjustment mechanisms extend carbon pricing beyond borders and level the playing field for domestic industry. Fossil fuel subsidy reform works on a similar principle and acts as a form of negative carbon pricing. Policy packages which combine elements of both quantity-based and price-based approaches are often implemented together.

Baseline-and-crediting mechanisms work by rewarding emission reductions relative to a defined baseline. A project which emits less than the set baseline can earn credits which are purchased by others who emit more than the baseline. There is no cap on the quantity of permissible emissions and no fixed price for the emission credits. The total emission reduction depends on the activity, and the price depends on the demand and supply of the credits. The Clean Development Mechanism (CDM) and Article 6.4 mechanism of the Paris Agreement are examples of this approach.

5.1.2 Development of Market Mechanisms

In 1990, the United States implemented the world's first large-scale market-based mechanism for lowering emissions. The SO_2 allowance trading program applied to power plants and was designed to reduce sulfur dioxide (SO_2) emissions, which were causing acid rain, under the Clean Air Act amendment. A fixed allowance to emit SO_2 was issued to power plants, and this cap was gradually lowered over time. Power plants that emitted lower than their allowance were permitted to sell their excess allowances to other plants. This led to a reduction of overall SO_2 emissions in a cost-effective way. It is reported that the program was highly successful and by 2007, annual SO_2 emissions had declined by 43% from 1990, despite more than a 26 percent increase in electricity generation from coal-fired power plants during the period [51].

Developed countries started experimenting with pilot ETSs, and the UK, Denmark, Norway and Sweden launched national pilots in the early 2000s. The EU ETS was launched in 2005, which is the world's largest multinational carbon market. This was followed by the US Regional Greenhouse Gas Initiative (RGGI) in 2009 and the California Cap-and-Trade Program in 2013.

The Kyoto Protocol introduced three flexible international market mechanisms in 1997. The International Emissions Trading (IET), the Clean Development Mechanism (CDM), and the Joint Implementation (JI) mechanism.

The IET mechanism allowed Annex I (developed) countries to trade emission allowances within themselves to meet their Kyoto targets cost-effectively. Under the cap-and-trade mechanism, each Annex I country had a binding emissions cap and received emission allowances. One Assigned Amount Unit (AAUs) represented one ton of CO_2 equivalent. A country emitting less than its cap earned surplus AAUs, while countries that emitted more than their cap had a deficit. Countries could buy and sell these AAUs, which led to the development of a global carbon market for allowances. As emission reductions were achieved where they were the cheapest, it lowered the total system cost.

The CDM is a form of baseline-and-credit mechanism which allowed an Annex I (developed) country to invest in projects which reduced emissions below a baseline in non-Annex I (developing) countries. This generated verified Certified Emission Reductions (CERs) where one CER represented one ton of CO_2 equivalent reduced. These standardized and fungible credits could be purchased via bilateral contracts and traded. Annex I countries bought the CER's to comply with their emission targets, thereby lowering their cost, compared to if emission reduction was undertaken domestically.

The JI mechanism allowed an Annex I (developed) country to invest in projects in another Annex I or Annex B country (which had a specific emission target) under the Kyoto Protocol. These projects generated Emission Reduction Units (ERUs), with one ERU representing one ton of CO_2 equivalent reduced (similar to CERs in CDM). These ERUs could be transferred and traded across Annex I countries via intermediaries or carbon markets. The major difference from the CDM was that this was a project-based crediting mechanism within a capped system. The mechanism effectively reallocated existing emission allowances across borders, allowing emissions reductions to be achieved where they were most cost-effective within Annex I countries. The JI mechanism later influenced the design of Article 6.2 carbon trading mechanisms between countries.

5.1.3 Market Design Principles

Several aspects are important while designing market-based mechanisms. Some of the key design features include having well-defined, measurable targets in terms of GHG emissions caps or reduction goals. Transparency is essential to build trust, while environmental integrity, which ensures that each emission reduction unit represents a real, additional, and permanent reduction in emissions, is non-negotiable for credibility. Flexibility measures like allowing banking and borrowing of credits must be built into the mechanism. Robust Monitoring, Reporting, and Verification (MRV) is essential for tracking and validating emission reduction. A liquid market provides effective price discovery, which is important for smooth trading. Regulatory certainty and long-term commitment provide confidence in markets and attract patient capital and investment. Principles of equity and fairness address distributional impacts and avoid disproportionate costs on vulnerable sectors or populations.

Although market mechanisms play an important role, these are not stand-alone but work in an environment that is supported by other policies. Overreliance on market mechanisms can lead to market failures due to information asymmetries, split incentives and behavioral barriers which limit the effectiveness of price signals. Other limitations include challenges in implementing them in sectors where there are distributed sources of emission which are difficult to measure and verify, such as land use, agriculture and landfills. Hence, market mechanisms need to be complemented by other policies including regulatory measures, standards, public investment, social support measures and informational measures.

5.2 Carbon Markets

Carbon markets facilitate reduction of GHG emissions at lower aggregate costs through trading of carbon credits. The system allows buying and selling of carbon credits, which represent one tonne of CO_2 removed or avoided. Compliance carbon markets function in jurisdictions where the government mandates a cap-and-trade program to limit emissions. In voluntary carbon markets, participants choose to offset their emissions by voluntarily buying carbon credits. Market-based approaches were discussed in Chap. 3 as a part of carbon pricing instruments, and some additional aspects are presented ahead.

5.2.1 Compliance Carbon Markets

A compliance carbon market is created and regulated by national governments or regional authorities. Specified entities are legally mandated to reduce emissions and face penalties for non-compliance. Such compliance markets have well-defined sectoral coverage, MRV rules and are closely governed by authorities in a specific jurisdiction. Compliance markets generally cover high-emitting industries such as power utilities, steel, chemicals, manufacturing industries and cement producers.

Regulators set a cap on total emissions and issue a limited number of permissible emission allowances. Companies must emit below their allowance limits or could trade the allowances if they have a surplus or deficit through an ETS. The EU ETS, California Cap-and-Trade (US), China ETS, and the UK ETS are some of the well-known compliance carbon markets. Global ETS revenues reached US$79 billion in 2025 and covered 14 $GtCO_2e$ (26% of global emissions), which makes it a relatively successful carbon trading mechanism [24]. Functional compliance markets are standardized, have high liquidity and credits are traded on a public exchange, leading to transparent pricing. At the end of 2025, 41 ETS were in force and 7 more countries are in the process of developing ETS.

5.2.2 Voluntary Carbon Markets

Market participants such as companies, organizations, or individuals can buy carbon credits from Voluntary Carbon Markets (VCM) to offset their emissions. These markets are not mandated by regulators but are governed by private standards and registries such as Verra, Gold Standard, American Carbon Registry, Climate Action Reserve and Puro.earth. They allow the fulfillment of corporate climate commitments by providing a mechanism to buy carbon credits to meet their carbon neutrality claims and also help to mobilize climate finance.

Project developers undertake projects for reducing or removing emissions in forestry, renewable energy, carbon removal, agriculture, methane capture, etc., and verified carbon credits generated from these projects are bought by corporates and institutions. These markets are more flexible and include a wide range of projects in different sectors. There are varying methodologies for certification and prices are decided through bilateral negotiations, brokers, exchanges, and on registry platforms. As there is no standardization and control on the quality of credits, it leads to fragmented and heterogeneous markets with a wide price range and low market liquidity. Voluntary carbon credits have also faced a backlash due to scrutiny over additionality, permanence, leakage, and double counting, leading to questions on the integrity of some of these credits.

Global VCM value is estimated to be about US$2.7 billion in 2025, which is a small fraction of the compliance carbon market [22]. An estimated 295 $MtCO_2e$ of carbon credits were issued in 2025, while about 200 $MtCO_2e$ credits were retired, leaving an annual surplus of about 90 $MtCO_2e$ [39]. Despite some challenges, voluntary carbon markets are forecasted to grow rapidly as they provide a flexible way for corporates to meet their climate commitments. They complement compliance carbon markets and facilitate private finance, broadening the demand for development of carbon reduction and removal projects.

5.2.3 Article 6 Mechanisms

The Paris Agreement introduced an international framework for countries to cooperate across borders for lowering the cost of GHG emission reduction. Article 6 mechanisms allow the transfer of carbon credits among countries, enabling them to meet their Nationally Determined Contributions (NDCs) cost effectively.

5.2.3.1 Article 6.2

Article 6.2 cooperative approaches include bilateral and multilateral mechanisms which allow countries to trade carbon credits directly with one another in a decentralized manner. Carbon credit units traded under Article 6.2 are known as Internationally Transferred Mitigation Outcomes (ITMOs) and represent GHG emission reductions or removals, which are measured in tonnes of CO_2 equivalent or other standardized metrics.

As this is a decentralized mechanism, governments can design their own programs, pricing and compliance trading systems. The host country can sell verified emission reductions to a buyer country, which can count these reductions against fulfillment of their NDCs. In return, the host country receives climate finance, technology transfer and capacity-building resources from the buyer country to implement decarbonization projects.

Clear rules for accounting and reporting are issued for Parties. After the host country sells an ITMO to the buyer country, it has to add the transferred GHG emissions to its domestic carbon inventory through a 'corresponding adjustment' to prevent double counting. Transactions and information sharing are done via the UNFCCC Centralized Accounting and Reporting Platform (CARP). Detailed reports are submitted by countries which are reviewed through the Article 6 Technical Expert Review (TER) process to ensure transparency and accuracy. Each ITMO is uniquely identified, and information such as country of issuance, year of issue and scope of authorization are recorded [56]. An international registry functions as an official record book and tracks ITMOs from issuance to use and automatically fills in required reporting information in an agreed electronic format (AEF).

ITMOs are a nascent market where agreements are signed bilaterally between governments. It is reported that 107 bilateral agreements were signed (till end March 2026), involving 63 different countries. 27.7 million ITMOs transactions have been recorded involving 5 different countries, which include E-mobility, clean cooking, forest conservation and waste projects. Major buyer countries include Japan, Singapore, Switzerland and South Korea. Seller countries include Ghana, Zambia, Senegal, Mongolia, Kenya and Indonesia among others. As the largest seller, Guyana transferred approximately 25 million ITMOs while Switzerland was the largest buyer with 13.6 thousand ITMOs [55]. Prices for ITMOs vary widely depending on several factors, and it was reported that ITMOs from Thailand, bought by Switzerland, were priced around US$28 per tCO_2e in late 2023 and early 2024, making it a cost-effective way for Switzerland to meet its NDC obligations [21].

5.2.3.2 Article 6.4

Article 6.4 established a new Paris Agreement Crediting Mechanism (PACM) to trade carbon credits in a UNFCCC-regulated centralized marketplace which would eventually replace the Kyoto Protocol's CDM. One Article 6.4 Emission Reduction (A6.4ER) equals one metric tonne of CO_2 equivalent avoided or removed. These are generated from projects in the power sector, which contributes to approximately half the number of projects, followed by buildings and industry. This mechanism allows private companies to directly develop emission-reduction or carbon-removal projects to earn credits which can be sold on a global marketplace. Five percent of the transaction proceeds from the sale of credits are routed to the UN Adaptation Fund. Two percent of all issued credits are automatically canceled upon issuance and cannot be used to offset emissions, which ensures an overall lowering of global emissions.

PACM framework is governed and overseen by the Article 6.4 Supervisory Body nominated by the UNFCCC, which approves carbon accounting methodologies, issues certificates and manages the centralized registry. Independent third-party auditors are accredited for validating projects and verifying the emission reduction before the issue of credits. Each host country is required to establish a domestic Designated National Authority (DNA) to screen projects and authorize the export of A6.4ERs. If a host nation authorizes the credits to be transferred abroad, a corresponding adjustment needs to be applied to its carbon inventory to prevent double counting, otherwise the emission reduction is counted towards the host's own NDC targets.

The mechanism is transitioning toward full operability, and there are stringent procedures covering baseline-setting, additionality, leakage, and non-permanence. The specialized PACM baseline methodologies for different projects are in the process of finalization. The centralized global Article 6.4 Mechanism registry is currently under development, which will eventually provide transparent tracking of cross-border transfers. It is expected that legacy carbon projects from the CDM will eventually become a part of the PACM pipeline once they comply with the new PACM methodologies and update their baselines.

As of the end of March 2026, 151 PACM activities had been approved by 28 different host Parties. Over 1130 new projects have been proposed and transition requests for about 1340 activities which were earlier under CDM have been received globally [55].

5.3 Energy Attribute Certificates

Electricity generated from renewable energy can be divided into two components, physical electrons and their associated environmental attributes. Energy Attribute Certificates (EAC) represent the environmental benefits associated with electricity generated from renewable energy sources such as solar, wind and hydro. Each EAC represents one megawatt-hour (MWh) of electricity generated and contains detailed data which is used to track the certificate. The digital certificate conveys information about the energy source used, point of origin, date and time of electricity generation and the reduction in carbon emissions. These environmental attributes are commodified and traded in specialized markets for the exchange of EACs. As the EAC represents the environmental attributes of electricity, companies can use it to claim use of renewable energy while continuing to buy physical electricity from their local grid, which may be generated from fossil fuels.

5.3.1 Renewable Energy Certificates

EAC is a generic term representing the environmental attributes of electricity generated from renewable energy sources. In North America (Canada and the US), these are known as Renewable Energy Certificates (RECs). There are two different market segments—compliance and voluntary markets.

In 1999, RECs were first introduced in Texas, followed by Arizona, California and Massachusetts in the early 2000s. The regulatory compliance system introduced Renewable Portfolio Standards (RPS), which mandated that a specific percentage of electricity sales for utility companies should come from renewable sources. This system is functional in over 29 U.S. states. To meet the requirements of the RPS, utilities can purchase RECs through specialized regional registries and trading platforms. Some of the platforms for tracking the transactions include the Western Renewable Energy Generation Information System (WREGIS) used by 14 states in Western North America and two Canadian provinces; Midwest Renewable Energy Tracking System (M-RETS) which operates across multiple states in the U.S. and Canada; and Pennsylvania, New Jersey, and Maryland Generation Attribute Tracking System (PJM GATS) used in several states in the Mid-Atlantic region.

Consumers can also choose to voluntarily buy RECs to make verifiable claims for corporate sustainability initiatives and to account for indirect emissions from purchased electricity (Scope 2 emissions). By voluntarily purchasing RECs, an organization can claim that the electricity it buys is carbon neutral, which is accepted under corporate standards such as the Greenhouse Gas Protocol. RECs bought voluntarily also support the company's sustainability reporting and Environmental, Social, and Governance (ESG) disclosures.

RECs generated from solar energy are known as Solar Renewable Energy Certificates (SRECs), which have a different compliance market in the U.S. Certain states have set a specific target, known as a 'solar carve-out', defining the minimum percentage of solar energy that must be sourced by electricity suppliers as part of their obligation under the state's RPS. The utility can produce this from in-state solar systems or through the purchase of SRECs.

The North American REC market is a patchwork of regional registries without an overarching national framework. Different states have different rules, and registries may not be interoperable. Due to this, the markets are fragmented, and the price of RECs varies in different states depending on supply, demand and applicable regulations. As an example, PJM Tier I REC prices reached $40/MWh in 2024 while prices for SRECs exhibited wide variation, reaching highs between $200 and $450 per MWh. The same platforms are used for the state-level compliance and the voluntary market, and measures are taken to avoid double-counting. It is reported that the RPS and the REC market have resulted in almost half of all growth in RE generation and capacity in the U.S. since 2000. This share has declined over time to 35% in 2023, but these market mechanisms continue to play a dominant role in driving RE growth [6].

5.3.2 Guarantees of Origin

Europe uses Guarantees of Origin (GOs) as a form of EACs. GOs are mandatory across the EU and some other European countries. They were introduced in 2011 alongside the EU's Renewable Energy Directive (RED). Each GO is a digital certificate which represents 1 MWh of renewable electricity and includes data on the renewable energy source (e.g., wind, solar, hydro), country and production source (e.g., name, location, commissioning date of the installation). They are issued (in each Member State of the EU) by designated national Issuing Bodies (IBs) which are members of the Association of Issuing Bodies (AIB). The GOs are managed by the inter-registry telecommunications hub, which facilitates the international exchange of GOs across countries according to a common set of rules and a linked system for tracking.

GOs are mainly used by utilities and energy suppliers to disclose the origin of electricity to consumers as it is legally mandated. GOs are also used in the voluntary market and companies use them to reduce scope 2 emissions and to meet ESG targets required by the Corporate Sustainability Reporting Directive (CSRD). GOs can be traded on spot auctions or through bilateral agreements. The prices are influenced by the type of GO (wind, solar, hydro), demand, and supply. In early 2023, GO prices peaked at around € 9–10 per MWh for hydro GOs, but prices corrected sharply in 2024–2025 with the average prices for generic GOs dropping to below €1/MWh.

5.3.3 Similar Energy Certificate Systems

Other countries have also adopted similar EAC systems. The UK Renewable Energy Guarantees of Origin (REGO) system separated from the EU GO system post Brexit. The J-Credits and Non-Fossil Certificates (NFCs) are two complementary national systems in Japan for tracking and trading environmental attributes. The J-Credits system certifies domestic GHG emission reductions from RE projects. The NFC system tracks and allows trade of environmental attributes of non-fossil electricity, specifically RE generated under the Feed-in Tariff (FiT) program.

China operates the Green Electricity Certificates (GECs) for trading the environmental attributes of RE. After a pilot program was started in 2017, this has become the world's largest EAC market and is now integrated with RE quota compliance. Korea Renewable Energy Certificates (K-RECs) and India's Renewable Energy Certificates are also used to meet the renewable purchase obligations in the compliance markets and in voluntary markets in national jurisdictions in these countries.

Australia has introduced two types of tradable certificates under the Renewable Energy Target scheme. Large-scale generation certificates (LGCs) are issued for large power stations and small-scale technology certificates (STCs) are issued for smaller renewable energy systems like rooftop solar, batteries, and solar water heaters. Designated entities are required to purchase and surrender certificates, which increases the demand for RE and stimulates investment. The primary market is complemented by a secondary market where financial institutions, traders, agents and installers can buy, sell and voluntarily surrender certificates [11].

5.3.4 International Renewable Energy Certificate

The International Renewable Energy Certificate (I-REC) system is a standardized framework for issuing, tracking, and redeeming Energy Attribute Certificates (EACs), which is administered by the International Tracking Standard Foundation (formerly known as the I-REC Standard Foundation).

The International Renewable Energy Certificate for Electricity, or I-REC(E), is a certificate that conveys information about the production of one MWh of renewable electricity injected into the grid. It mirrors the GOs in Europe and RECs in North America and is adopted in more than 60 countries across Asia, Oceania, Africa and South America. I-REC(E) is used by corporates to meet their voluntary RE targets and is recognized by global corporate sustainability reporting frameworks, such as the CDP (formerly Carbon Disclosure Project) and Global Reporting Initiative (GRI). The system enables cross-border trading of certificates and is useful in countries without domestic REC frameworks.

5.3.5 24/7 Carbon-Free Energy Certificates

24/7 Carbon-Free Energy (CFE) certificates guarantee that a kilowatt-hour (kWh) of electricity was generated from a carbon-free source, time-matched to a specific hour of consumption. These are also called time-based energy attribute certificates (T-EACs) or granular certificates (GCs). 24/7 CFE EACs are procured on an hourly basis to match consumption with the generation of carbon-free energy on the same local or regional grid to lower local emissions. Unlike traditional EACs that can be purchased on a yearly basis, 24/7 CFE EACs address the volume and temporal alignment between demand and supply, ensuring that the energy procured is matched temporally to RE production. As there is an hourly level of granularity built into the system, companies can claim to operate on 100% carbon-free electricity, 24 h per day, 7 days a week, rather than claiming to use renewable energy over an annual time period.

The global market for 24/7 CFE EACs is still evolving, and there is no centralized global marketplace as yet. Procurement typically involves bilateral agreements with energy firms like EDF, AES, Statkraft, and ENGIE and corporate buyers like Google, which have set an ambitious goal to operate on 24/7 carbon-free energy by 2030 [23]. 24/7 CFE EACs encourage investments in clean energy infrastructure by creating stronger financial incentives for RE developers to build projects that deliver 24/7 carbon-free energy.

5.4 Markets for Resource Efficiency and Environment Protection

Market-based approaches can be used to encourage the efficient allocation and use of resources. They create financial incentives for adoption of resource-efficient practices to minimize environmental impacts. They address market failures by internalizing the environmental cost of resource extraction and waste disposal, which makes resource-intensive products or wasteful activities more expensive, providing a price signal to invest in resource-efficient technologies, products, and services.

5.4.1 Energy Efficiency Certificates

Energy efficiency certificates (also known as white certificates or energy saving certificates) are tradable instruments for encouraging adoption of energy saving technologies and practices. The system provides financial incentives to industries and businesses to save energy. Energy savings incurred through installing energy-efficient equipment (e.g., energy-efficient lighting, improved building insulation) are verified by an Energy Service Company (ESCO) that monitors the performance of the new system and energy efficiency certificates are issued. These certificates can then be bought by obligated entities such as utility companies on a digital marketplace or through bilateral contracts to meet their shortfall in achieving the energy efficiency obligation quotas.

These tradable certificates were first made operational in Italy in 2005, followed by France in 2006, where they are known as Certificats d'économies d'énergie (CEE). Australia, India, Poland, Spain, the UK, the US and some other countries have adopted different versions of energy efficiency certificates. An energy saving certificate has different units of measurement. One certificate in Italy or Poland may represent energy savings of one tonne of oil equivalent (toe), while in Australia, a certificate represents one nominal MWh of energy saved. In France, the CEE is expressed in kWh cumac (cumulated and discounted), which measures energy savings over the entire lifetime of equipment, discounted to present value.

The major benefit, apart from energy savings, which translate into avoided CO_2 emissions, is the additional revenue stream from selling these certificates. This offsets the costs of energy efficiency investments and leads to a shortened payback period. The mechanism also offers flexibility to the obligated parties, which can meet their energy saving targets through their own investments or through the purchase of certificates in a cost-effective manner. It also encourages innovation for the development of new energy-saving technologies and fast-tracks technology adoption. Challenges include high costs of verification and monitoring, price volatility in certificates affecting revenue streams and insufficient market liquidity. Despite this, energy efficiency certificates are emerging as important tools to meet emissions reduction targets and are increasingly being adopted.

5.4.2 Circular Economy Credits

Circular economy credits are market-based instruments that represent the verified collection and recycling of waste material such as plastic, glass or electronics, avoiding their disposal into the environment.

5.4.2.1 Attributes of Recycled Content

Attributes of Recycled Content (ARCs) are environmental commodities generated under the Recycled Material Standard (RMS) program [46]. It serves as a voluntary, market-based tool to increase recycled content in packaging. Currently, only entities located in North America are eligible for certification to the RMS. Each ARC is a tradable certificate that represents the environmental benefits of producing one metric ton of recycled output. The physical flow of the recycled material is separated from the certificate. ARCs are issued to recyclers after a certification body audits a new facility and certifies that it meets the 'additionality' criteria. Companies can purchase ARCs to meet their voluntary recycled content commitments and the revenue from the sale of ARCs goes to recyclers for financing infrastructure investments. An online registry system maintains transparency and tracks each ARC.

5.4.2.2 Extended Producer Responsibility Certificates

Extended Producer Responsibility (EPR) certificates and credit trading are a market-based mechanism to promote a circular economy. It creates a market for recycled materials and allows producers to meet their recycling and waste management obligations by purchasing EPR certificates from authorized recyclers. The mechanism works alongside an environmental compliance framework such as the one in India, which mandates responsible waste management and is supported by several environmental regulations, including the plastic waste management rules, E-waste management rules, battery waste management rules, and hazardous waste rules. The Central Pollution Control Board (CPCB) is the apex regulatory body for EPR management in India and sets mandatory annual targets for the collection and recycling of waste streams such as plastic packaging, e-waste, batteries and tires for the producers, importers and brand owners (PIBOs) [50]. The EPR certificate trading system has recycling companies which act as sellers while producers and importers of goods are obligated to buy the EPRs if they do not meet their recycling targets. Registered recycling companies collect and process plastics, e-waste, batteries and tyres. After the waste is recycled using environmentally sound management standards, a digital EPR certificate specifying the type and quantity of material processed is issued. The certificates are verified by the regulator and uploaded on an online portal. Producers with a compliance shortfall can purchase the surplus EPR credits from the CPCB portal and can use them as verifiable proof of compliance. The mechanism offers a flexible way to avoid penalties for non-compliance for buyers while recyclers generate revenue from selling the EPR certificates, which incentivizes the development of waste management infrastructure.

5.4.2.3 Plastic Credit Certificate

A plastic credit certificate is a transferable unit representing a specific quantity of plastic that is avoided from use, collected and managed, or recycled, as a result of a project activity [60]. The tradable certificate can be issued by one party and sold to another through a transaction chain. The certificate represents that a verified quantity (1 kg or 1 ton) of plastic waste has been collected, processed, or recycled. Projects that involve household collection and management of plastic, recovery of plastic waste in landfills and management through co-processing, mechanical or chemical recycling of plastic waste into secondary raw material can generate plastic credits. A project developer can design plastic recycling projects that are validated by an auditor. Once the project is registered and the plastic is recycled, the auditor verifies and issues the plastic credits. These credits can then be sold to buyers directly or through a trader on an online platform. After the buyer purchases the credit, it can claim the environmental benefits of the plastic credit, which is then retired from circulation. The proceeds of the sale of plastic credit become an additional revenue stream for the project developer, which encourages plastic recycling.

Some of the commonly used programs with publicly available standards, methodologies and registries include Verra Plastic Waste Reduction Standard (PWRS), Plastic Pollution Reduction Standard (PPRS), GreenBlue Recycled Material Standard (RMS) and Zero Plastic Ocean-Bound Plastic (OBP). Plastic credits can be bought and sold on platforms such as Plastic Credit Exchange (PCX). Currently, there is no globally accepted mechanism for trading plastic credits and transfer of ownership is only possible once between the buyer and the seller [43]. As of December 2023, approximately 160 projects were listed on plastic credit registries, mainly in developing countries across East Asia and the Pacific, and South Asia [60]. Plastic credit prices range from US$140–670 per credit (equivalent to one ton of plastic waste) depending on the type of waste management activity, project location, and the complexity of the certification process. The main buyers of plastic credit are plastic producers, Fast-Moving Consumer Goods (FMCG) companies, pharmaceutical companies, personal care brands and nonprofit organizations. Although plastic crediting is largely a voluntary market, it can be used as a compliance mechanism within EPR schemes where producers can purchase plastic credit to meet their mandated obligations, such as in India, the Philippines and the UK.

5.4.3 Circular Credits Mechanism

Circular Credits Mechanism (CCM) is a market-based mechanism created by BVRio (Bolsa Verde do Rio de Janeiro—Environmental Exchange of Rio de Janeiro) to incentivize waste collection, recovery, recycling and sorting through effectively subcontracting waste-recovery services, thereby helping companies mitigate their waste footprint [10]. It enables companies to buy circular credits from sellers such as waste-management projects and waste-pickers associations.

After the project registration, where the waste-management activity and the expected benefits are defined, the project self-declares expected environmental results and the platform allocates an estimated number of circular credits to the project. The project carries out the waste-recovery activities (collection, sorting, transport to recycler/destination) and tracks the relevant data such as weight recovered, source, and destination. When a buyer expresses interest, the *Circular Action Hub* facilitates negotiation and transaction, acting as a marketplace and registry. After the activities are implemented, an independent third-party auditor verifies results and compliance with the CCM standard and the verified credits are issued and transferred to the buyer. Once credits are used for corporate claims such as waste-footprint mitigation, they are retired in the registry.

CCM uses an inverted project cycle where projects can advertise self-declared offers before the project is verified. This lowers entry barriers for Small and Medium-sized Enterprises (SMEs) or waste-picker groups that may not have upfront resources for registration or audit costs. The mechanism creates a revenue stream for waste-workers and hence is unique as it combines environmental objectives with social inclusivity. CCM targets waste recovery and recycling, not reduction or prevention of waste generation and hence it must be complemented by other mechanisms to reduce waste generation.

5.5 Market Mechanisms Based on Natural Capital

Certain market mechanisms have been developed to trade in ecosystem services derived from natural capital. A monetary value can be assigned to ecosystem services such as improved biodiversity, flood risk reduction, improved water quality and reduction in nutrient runoff through developing projects which conserve and restore nature. Natural capital markets use a cap-and-trade approach to generate verifiable environmental credits that can be sold and bought. Typically, a project developer implements projects such as wetland restoration or reforestation which has measurable environmental benefits. An independent, third-party verifies the benefits from the project (example, measuring an increase in species) based on established methodologies and credits are issued, where a credit represents a specific unit of environmental improvement. Companies can purchase these credits to meet their voluntary ESG commitments or regulatory requirements. This allows mobilization of private capital into activities for conservation and restoration of nature while offering a financial incentive for landowners to protect and undertake ecosystem-enhancing activities.

5.5.1 Water Quality Trading

Water Quality Trading (WQT) is a market-based approach for improving local water quality. It aims to control the discharge of pollutants such as nitrogen and phosphorus from multiple sources into waterways and can achieve higher water quality standards at lower abatement cost than traditional engineered solutions [58]. Water quality is treated as a tradable commodity and the system functions when accompanied by stringent regulatory standards. Polluters can discharge pollutants up to a certain limit in the waterway. Entities can earn water quality credits by discharging below this limit, and the holder of the credits can sell any unwanted credits to another user. Hence, the mechanism allows facilities to trade pollution-reduction credits from other facilities in the same watershed or from non-point sources (farms and landowners), to meet overall regulatory requirements for water quality. Water quality trading credits are measured in units of pollutant load reduced, such as pounds of nitrogen or phosphorus per year, below a required baseline.

WQT can provide flexibility to entities for investing in water treatment technologies or buying credits, whichever reduces their overall compliance costs. Farmers can voluntarily participate in the scheme to implement agricultural practices (e.g., streamside buffers, conservation tillage) to control agricultural runoff of nutrients within the watershed and receive monetary benefits for reducing pollutants. WQT also provides environmental co-benefits such as stabilizing streambanks, providing wildlife habitat and improving biodiversity.

More than 50 water quality trading programs have been identified worldwide, of which 26 are active and 21 are under consideration or development. Most of these are in the United States, with four in Australia and one each in New Zealand and Canada [18]. In 2012, three U.S. states, Ohio, Indiana and Kentucky, developed a regional interstate water quality trading framework in the Ohio River basin, which is now the world's largest WQT program. In 2014, pilot trades were conducted which represented the first interstate credit transactions for water nutrients in the United States [16].

Water credits are generated by projects which reduce the volume of fresh water used for a specific purpose (as compared to a baseline) and improve the quality of water discharged into the water body [1]. Credit Aqua Positive (CAP) is a form of positive water credit which represents the impact achieved through projects that help preserve and restore water resources. One CAP equals 1,000 m^3 of saved water footprint, including the environmental and social benefits of water footprint compensation projects [2]. The Act4Water standard provides a regulatory framework and a voluntary CAP market, which allows projects to earn a certain number of CAPs. Each CAP is priced based on the investment needed to carry out and maintain the project. Organizations can offset their water impact by buying CAPs, and the proceeds fund projects to support water-positive outcomes.

5.5.2 Regulated Biodiversity Markets

Regulated biodiversity markets are designed to compensate for the unavoidable damage to biodiversity caused by development projects. The government limits the allowable harm to certain listed endangered species and habitats, such as through implementing the 'Endangered Species Act' in the U.S. When a project has a negative environmental impact, the project developer must take steps to avoid, minimize or mitigate the impact of the project. This must be undertaken as per the established mitigation hierarchy that mandates project developers to avoid and minimize their negative environmental impacts as much as possible on-site. But if the impacts cannot be fully mitigated at the site, the residual impacts must be compensated by purchasing biodiversity credits. Regulated parties can buy these credits from private conservation banks that implement projects which restore habitats for endangered species. Conditions often include a matching of projects to ensure that the credits are 'like-for-like' and 'local-to-local' to ensure ecological equivalence. Biodiversity credits provide a financial incentive to protect endangered species' habitats permanently and channel a consistent flow of private capital into conservation activities. These credits can be stored in mitigation banks, which can be used to offset the environmental impacts of projects.

One example is the UK's Biodiversity Net Gain (BNG) policy, which mandates that project developers must ensure a minimum 10% measurable increase in biodiversity compared to the pre-development state, which creates a market for biodiversity credits. A standardized biodiversity metric is used to calculate the value of habitats before and after project development. Specialized habitat banks undertake long-term conservation or restoration projects which generate measurable Biodiversity Units (BU), which are verified by an independent third party and registered in a transparent public registry. Developers can purchase these credits to meet their legal obligations and a regulatory body ensures that the conservation projects deliver the promised ecological benefits. The price is a function of market dynamics and is impacted by the costs of project management and the type of projects. Another example is the New South Wales Biodiversity Offsets Scheme in Australia, which uses Biodiversity Stewardship Agreements (BSAs) to create a biodiversity market mechanism [40].

5.5.3 *Habitat and Species Banking*

The primary goal of habitat and species banking is to offset habitat and species loss, especially in endangered species. In a habitat banking system, specific areas of land are permanently protected, restored, or enhanced to generate biodiversity credits. The focus is on protecting a specific type of ecosystem like wetlands, woodlands and riparian zones. Species banking or conservation banking is a form of habitat banking specifically designed to protect and manage habitats for plant and animal species that are endangered or threatened. Credits are generated based on the projected ecological benefits of the restored habitat, which are purchased by developers as they are regulated under species protection laws such as the 'Endangered Species Act'. The revenue from the sale of credits is used to implement the long-term monitoring and maintenance of the habitat or species bank.

5.5.4 *Wetland and Stream Mitigation Banking*

Wetland and stream mitigation banking is a system to offset loss or degradation of wetlands and streams. Credits are generated by restoration, creation, or preservation of wetlands and streams in a specific location to create mitigation banks. These credits are then sold to offset unavoidable harm to hydrological and water quality degradation from development projects undertaken in a different location. The driver for buying offsets in the U.S. is the regulation under the 'Clean Water Act' and other federal laws. As the credits are released for sale only after successful verification of results by projects implemented in the mitigation bank, confirmed ecological benefits are provided upfront.

5.5.5 Voluntary Biodiversity Credits

Unlike the regulated market, Voluntary Biodiversity Credits (VBCs) are optional and can be purchased by any organization to meet specific biodiversity goals that contribute to nature restoration. VBCs are not offsets and they don't compensate for the buyer's environmental harms elsewhere. Companies use VBCs as a voluntary contribution to nature conservation or for philanthropic purposes. World Wildlife Fund (WWF) defines a biodiversity credit as "a certificate that represents a measured and evidence-based positive biodiversity outcome from nature restoration, conservation or stewardship activities, that is not purchased for the purpose of offsetting residual negative impacts caused to biodiversity". A VBC must achieve and measure demonstrated biodiversity outcomes in terms of avoided loss or biodiversity gain. VBCs are relatively new, and standards are evolving. Some projects which generate wildlife credits are being piloted by WWF in Tanzania [61] and to conserve forests in France [62]. About 30 developers are providing VBCs and the most developed schemes currently cover more than 800,000 hectares with US$8 million in financial commitments [28]. As there are no global market regulations and supervision of the VBCs, the secondary market for trading of VBCs is limited.

5.5.6 Challenges in Natural Capital Markets

Like other market mechanisms, there are challenges related to proving additionality and ensuring that the environmental gains are permanent. Establishing a baseline is difficult and as measuring biodiversity is inherently difficult, creating universally acceptable standardized metrics is challenging. Governance, enforcement and implementation capacity is at best limited, which creates fragmentation of markets. Reducing ecological complexity to a single metric of biodiversity units may lead to species-level and ecosystem functional losses. Concerns on greenwashing need to be addressed as offsets allow companies to continue environmentally damaging activities. Social issues such as fair and equitable distribution of benefits, particularly regarding Indigenous communities and local land rights, often remain under-addressed.

5.6 Nature-Based and Farming Markets

5.6.1 Forestry Markets

5.6.1.1 Agroforestry Credits

Agroforestry is a land management practice that combines the planting and management of trees and shrubs with crops and pasture, providing benefits of improved soil health, crop yields, resilience to climate change and biodiversity [9]. Agroforestry projects sequester atmospheric carbon dioxide and can generate carbon credits. Methodologies such as Verified Carbon Standard (VCS) and the Climate, Community and Biodiversity (CCB) standards are used to calculate carbon sequestered from agroforestry projects, which are certified by independent bodies such as Verra, Gold Standard and American Carbon Registry. The issued credits are then traded in voluntary and compliance carbon markets and generate an income stream for project developers and farmers. Agroforestry projects and credits also lead to environmental benefits such as enhanced soil health and water quality and have social co-benefits such as promoting livelihoods.

5.6.1.2 REDD + Carbon Credits

Forests store a large amount of carbon as biomass, which can be released through deforestation. Reducing emissions from deforestation and forest degradation (REDD) + carbon credits are generated from activities which result in avoided deforestation and forest degradation. The original REDD program focused only on stopping deforestation and forest degradation. The ' + ' expanded the scope of REDD activities to include five types of projects which can generate REDD + credits [48].

- Reducing deforestation through preventing conversion of forest land to agriculture, mining, or development projects
- Reducing forest degradation by limiting damage to forests from unsustainable logging or fire
- Conservation of forests in threatened areas
- Sustainable forestry management practices through harvesting timber and other products sustainably
- Restoring degraded forests.

REDD + carbon credits can be sold in voluntary or compliance carbon markets, and projects can earn an additional revenue stream by protecting forests. For evaluating the baseline, project developers identify historical deforestation rates to calculate the reference point for emission reductions. The forest cover is tracked over time, and the impact of the project is compared against the baseline. Third-party auditors issue credits after verification of the measurements and emission calculations. Each verified emission reduction of one ton of CO_2 avoided becomes one carbon credit and is recorded in independent registries like Verra or Gold Standard, from where they can be purchased. The price of these credits varies significantly from US$5–20 per ton and is a function of the project quality. Challenges for REDD + projects include baseline measurement, underreporting of deforestation, and permanence risk due to forest loss and MRV issues.

Jurisdictional REDD + credits are emerging as an alternative to project-level REDD + credits. In this approach, the entire forest in a national (country) or subnational (e.g., state or province) jurisdiction must be considered when setting a baseline. This leads to reduced risk of inflated baselines and over-issuance of credits. Monitoring deforestation across a whole jurisdiction also prevents carbon leakage, as large-scale monitoring is able to detect if deforestation moves from inside the project area to an area that is not being monitored. MRV for large areas is cost-effective, and as these programs involve entire countries, there is an incentive for the government to develop policies and regulations beyond project-based REDD+ [54].

5.6.1.3 Peatland Credits

Peatlands are terrestrial wetland ecosystems in which waterlogged conditions prevent plant material from fully decomposing. The production of organic matter in peatlands exceeds their decomposition rates, which makes peatlands large carbon sinks [31]. Peatlands store up to one-third of the world's soil carbon, more than all forests globally. However, when they are developed for agriculture, the exposed peat decomposes, releasing CO_2 emissions. Projects that restore the water table and revegetation can earn carbon credits as they restore and conserve peatland ecosystems, which avoids further CO_2 emissions. These projects can be registered under the VCS or Gold Standard, which certify and track the amount of carbon stored or avoided. Regional standards, such as the 'IUCN Peatland Code' in the UK, also provide certification standards suited to local conditions [36]. These credits can be sold in the VCM, where corporations can purchase them to offset their emissions.

5.6.2 Sustainable Land Management

Sustainable land management (SLM) is defined as "the use of land resources, including soils, water, animals and plants, for the production of goods to meet changing human needs, while simultaneously ensuring the long-term productive potential of these resources and the maintenance of their environmental functions" [59]. SLM can provide three types of carbon benefits: carbon storage in natural forests, grasslands, and wetlands, carbon sequestration where biomass growth removes carbon from the atmosphere and stores it in the form of soil organic carbon (SOC); and through lowering GHG emissions from agricultural production, including from land use change.

5.6.2.1 Soil Carbon Credits

Soil Carbon Credits (SCC) are credits generated from agricultural practices to increase SOC sequestration, reduction and avoidance of GHG emissions [13]. Farming practices like lowering the use of fertilizers, reducing tillage, increasing production of cover crops, improving residue management of crop waste, crop rotation and agroforestry help in increasing the amount of carbon stored in the soil or reducing emissions from farming. These activities also improve soil fertility and water retention.

Grasslands and rangelands that are grazed by domestic or wild animals can also act as a significant carbon sink as they store and sequester carbon. Approaches which protect grasslands and avoid further conversion for agriculture, rotational grazing, improved grazing management, prescribed burning and restoration of degraded grasslands can enhance carbon storage capacity in grasslands. Independent organizations like Climate Action Reserve use the U.S. Grassland Project Protocol to verify and certify grassland carbon projects [12]. After third-party verification of projects, project developers can earn carbon credits based on sequestered carbon. These credits can then be sold on the VCM, bringing additional revenue for farmers. VM0042 methodology for improved agricultural land management by Verified Carbon Standard and soil carbon method 2021 by Australian Carbon Credit Unit (ACCU) Scheme are some of the standards used for MRV of SCC.

5.6.2.2 Reef Credit Scheme

The reef credit scheme is a market-based approach to improve the quality of water entering the Great Barrier Reef in Australia [15]. It works similarly to carbon credits and puts a monetary value on an ecosystem service. A 'reef credit' represents a quantifiable, verified volume of pollutants such as sediment, pesticides and dissolved inorganic nitrogen which have been removed from water flowing out to the reef. Reef credits are unique as they target land management practices to protect corals in the sea.

Grazing management approaches, which incentivize sustainable grazing practices to reduce sediment runoff and nutrient pollution for better water quality in the Great Barrier Reef catchment area, have been introduced. A Grazing Land Management (GLM) methodology has been developed under the 'Reef Credit Standard' which includes measures for pasture management, soil erosion control, and reduced sediment and nutrient runoff into waterways that feed the Great Barrier Reef. The GLM method quantifies and values the reduction in sediment losses resulting from improved grazing land practices and aims to incentivize landholders to achieve a high level of ground vegetation cover before high-intensity rainfall, to reduce sediment run-off onto the reef. Unlike carbon, the pollutant being controlled is the fine sediment. The developed methodology uses the Revised Universal Soil Loss Equation (RUSLE) and Dynamic Reference Cover Modeling (DRCM) to provide a measurable approach for fine sediment abatement. Landholders undertake projects to manage grazing lands which cut pollutant flow. Projects are verified by independent third parties and reef credits are issued. *Eco-Markets Australia* manages the registry and sells the reef credits to companies to support their claims of supporting reef health [14].

5.6.2.3 Livestock and Manure Management

Globally, livestock production contributes to about 15% of all anthropogenic GHG emissions, of which about half comes from methane. Livestock and manure management carbon credits can be earned by adopting practices to reduce or capture methane emissions. These can be sold on voluntary and compliance carbon markets to provide a new revenue stream for farmers.

There are two existing methodologies for certifying carbon credits: reducing enteric methane emissions from animal digestive processes (enteric fermentation) and reducing methane emissions from livestock manure management. Enteric methane reduction solutions that qualify under Verra, Gold Standard, and CAR programs include feed additives and diet improvement, genetic improvements, herd management and improving animal health. 'Methodology for quantification of GHG emissions reductions from improved management in smallholder dairy production systems using a Standardized Baseline' (developed by Gold Standard) and 'Methodology for the reduction of enteric methane emissions from ruminants through the use of feed ingredients' (developed by Verra) are some of the methodologies which outline the specific procedures and methods for implementing, measuring, monitoring, reporting and verifying project activities [17].

Manure stored in anaerobic conditions creates methane. Manure management activities that decrease the amount of time that manure spends in low-oxygen conditions reduces the amount of methane emitted. Manure handling strategies such as biogas capture and upgrading for energy generation, biogas capture and upgrading for fuel production, biogas capture and destruction through flaring, or preventing the creation of biogas through solid–liquid separation or other aerobic manure management techniques are some of the activities that can be used to generate carbon credits. Individual animal farms can generate carbon credits through implementing livestock and manure management practices. Project developers help by quantifying the baseline emissions and reduced emissions resulting from the intervention, which are verified. The issued carbon credits can be sold on the registry, and the revenue flows back to farms and project developers.

5.7 Carbon Dioxide Removal Credits

Carbon Dioxide Removal (CDR) is a method for extracting CO_2 from the atmosphere and storing it in geological formations, land, oceans, or durable products. Natural CDR involves nature-based approaches such as afforestation, soil carbon sequestration, and blue carbon methods, while technological CDR includes Direct Air Capture (DAC), and enhanced mineralization. High-quality CDR credits are permanent and keep CO_2 stored for centuries preventing it from being released back into the atmosphere.

5.7.1 *Direct Air Capture Credits*

Carbon credits can be issued to projects that remove carbon using Direct Air Capture (DAC) technologies, which use liquid solvent and solid sorbent systems. The chemical process includes capturing CO_2 from air using a liquid solution or a solid filter that bonds with the CO_2 molecule. After the absorption cycle is completed, the material is heated to a high temperature, which breaks the chemical bonds and releases the CO_2 as a concentrated gas stream with 95–99% purity. The purified CO_2 gas is compressed to a supercritical state and is transported through pipelines to permanent underground storage sites or delivered for making products such as cement and synthetic fuel.

Permanent geological storage can be undertaken by injecting compressed CO_2, 1–3 km underground into porous rock formations. The CO_2 gradually reacts with surrounding rock to form solid carbonate minerals, which has minimal reversal risk. Captured CO_2 can also be combined with hydrogen to synthesize carbon–neutral jet fuel or diesel. On burning, the captured CO_2 is released, which creates a closed carbon loop. As this is not permanent removal, it does not lower atmospheric CO_2 concentrations, but provides carbon–neutral energy. CO_2 can also be used to produce building materials such as concrete, plastics, and carbon fiber, which can store the captured carbon in usable material for decades.

One carbon removal credit represents one ton of CO_2 verifiably removed and stored directly from air. Unlike carbon capture technologies for point sources of emissions such as power plants or cement factories where CO_2 concentrations are around 3–20%, DAC technologies work with standard atmospheric air containing about 0.042% CO_2. Thus, DAC plants can be built where there is cheap RE or adjacent to geological CO_2 storage sites [49]. The cost of capturing carbon through DAC is around \$600–\$1,000 per ton of CO_2 removed, which is 10–20 times more expensive than capturing CO_2 from point sources such as a power plant smokestack. The technology requires 1–5 tons of water per ton of CO_2 captured for cooling, which is also a major concern in areas with high water stress.

Some existing plants include the 'Mammoth' plant in Iceland by Climeworks, which has a capacity to capture 36,000 tons of CO_2 per year. Occidental Petroleum's Stratos facility is under construction in Texas with a planned capacity of 500,000 tons annually. Over 100 other projects are in various development stages globally. These projects are certified by third-party standards like Verra, Gold Standard and Isometric [35]. Some of the largest DAC projects with their capacity are shown in Table 1.

Table 5.1 DAC projects with location, certification and capacity

Project name	Location	Certification	Annual removal (tons CO_2)
STRATOS (1PointFive)	Texas, USA	Verra (expected)	500,000 +
Project Bison (CarbonCapture)	Wyoming, USA	Puro.earth	200,000 + (2025–2026)
Deep Sky Alberta	Alberta, Canada	In development	10,000 + (scaling)
Climeworks Iceland	Iceland	Gold Standard	36,000 +
Heirloom California	California, USA	Verra	1,000 + (pilot)

Data source [45]

The main buyers of DAC credits is Microsoft, which agreed to buy 500,000 metric tons of credits from 1PointFive and smaller amounts from Project Bison and Deep Sky. JP Morgan and Palo Alto Networks also purchased carbon removal credits from DAC.

5.7.2 Biochar Credits

Production of biochar is a source of CDR credits, and each credit represents one metric ton CO_2e of verified and sequestered carbon. Plants absorb CO_2 as they grow and release the CO_2 back into the air as they decompose or burn. When the organic material is heated at high temperature in the absence of oxygen, the material does not burn, but its complex molecules break down to produce syngas, pyrolysis oil and biochar, which is a solid, carbon-rich material. This thermal decomposition of organic materials is called pyrolysis and converts up to 50% of the plant's carbon into a highly stable form. Biochar, when applied to soil or used in construction materials, effectively locks away carbon for long term in a durable form. Using biochar also has several co-benefits like enhancing soil fertility, nutrient retention and reducing emissions of nitrous oxide and methane from soil.

Biochar credits use standards such as 'Verra VM0044—Biochar utilization methodology', 'Puro.earth—biochar standard' and 'Isometric biochar methodology', which define the eligible feedstocks, pyrolysis conditions, and MRV requirements. Registries like Verra, Puro.earth, Isometric or Carbon Standards International (CSI) certify the projects and issue the credits after auditing their additionality, durability, and measurability. Once certified, biochar credits are listed on registries and sold to corporates, investors, or governments seeking to offset emissions or to meet net zero goals. Apart from the revenue stream from the credit, biochar producers can also earn revenues from the sale of physical biochar as compost or as an additive to soil [4].

The price of biochar credits varies between US$100–$250 per ton CO_2e depending on feedstock, production technology for biochar (temperature, pyrolysis conditions), and how it is stored or used. Biochar credits are bought by corporates such as Microsoft, Google and Swiss Re and accounted for more than 90 percent of durable CDR delivered in 2023–2024. It is estimated that the global demand for durable carbon removal is expected to hit 40–200 $MtCO_2$ each year by 2030, which would require scaling up their supply [8].

5.7.3 Enhanced Rock Weathering Credits

Rock weathering is a natural process which is estimated to sequester about 1 Gt of atmospheric CO_2 per year. During this weathering process, atmospheric CO_2 dissolves in rain droplets or river water and forms weak carbonic acid (H_2CO_3), which reacts with alkaline minerals in silicate and carbonate rocks, generating stable bicarbonate (HCO_3^-) and carbonate (CO_3^{2-}) ions. The dissolved inorganic carbon precipitates as minerals in the soil are transported via rivers to the ocean, where they are locked in that state for a long duration.

Enhanced Rock Weathering (ERW) credits are a type of carbon removal credit generated by projects that accelerate the natural process of rock weathering to permanently sequester atmospheric carbon dioxide. One ERW credit represents a verified metric tonne of CO_2 that has been durably removed from the atmosphere. ERW has high permanence, as the storage of carbon as bicarbonate in the ocean is considered highly durable over centuries. However, the issuance of credits requires rigorous MRV processes. ERW involves different steps to speed up the natural rock weathering process. Rocks having minerals with high weathering reactivity and cation content (mafic and ultramafic rocks and rocks containing silicate minerals such as basalt) are mined and ground into a fine powder to increase their surface area. The crushed material is then spread onto agricultural land. When rainwater containing dissolved CO_2 comes into contact with the highly reactive rock dust, CO_2 is converted into stable bicarbonate and carbonate ions, preventing the CO_2 from returning to the atmosphere. ERW has other agricultural co-benefits such as improved soil health due to the addition of nutrients like potassium, magnesium, and calcium in the soil.

The first ERW carbon removal credits were verified and issued by Isometric for a project by InPlanet in Brazil in early 2025. These credits are bought by the digital payments company Adyen, which will use them to offset their own residual emissions [25]. Other companies like Puro.earth are also developing ERW carbon crediting methodology [44].

5.7.4 Bioenergy with Carbon Capture and Storage Credits

Bioenergy with Carbon Capture and Storage (BECCS) credits are another form of CDR credit and represent one metric tonne of CO_2 that has been captured from biomass utilization and permanently stored underground or utilized in long-term products. BECCS involves capturing CO_2 from biogenic sources such as rice straw, sugarcane, agricultural residues, crop waste and municipal solid waste. In April 2025, Microsoft signed a carbon removal agreement with AtmosClear to buy 6.75 million metric tons of BECCS credit over a 15-year period from a new BECCS facility being developed in Louisiana. Construction of the US$800 million plant, which will capture 680,000 metric tons of CO_2 annually using sugarcane waste and forest trimmings, will commence in 2026 and is expected to become operational in 2029 [26].

5.7.5 Marine Carbon Dioxide Removal

Marine Carbon Dioxide Removal (mCDR) involves the use of technologies to enhance the uptake of carbon dioxide from the ocean. It involves carbon removal from the surface of the ocean and storing it for long periods of time, allowing the ocean to absorb more atmospheric CO_2. It can be categorized into biotic and abiotic approaches. Biotic or biological pathways rely on photosynthesis by microalgae in phytoplankton, seaweed and kelp to absorb CO_2. Abiotic or chemical pathways rely on chemistry to increase the amount of atmospheric CO_2 dissolved in seawater, where it can be stored for thousands of years.

5.7.5.1 Biological Approaches

Restoration, expansion, and protection of coastal marine ecosystems like mangroves, salt marshes, and seagrasses are some ways to capture and sequester CO_2. Seagrass meadows absorb CO_2 from the water column through photosynthesis and sequester carbon through the uptake and storage of CO_2 in their leaves and roots. Mangroves also sequester carbon through organic matter accumulation in their roots and sediment. mCDR projects focus on protecting existing mangroves from deforestation and coastal development as well as cultivating seagrass or seaweed in underwater plots. Restoration activities also include planting mangrove saplings in degraded areas.

A blue carbon credit represents carbon dioxide removed or stored by coastal and marine ecosystems. Blue carbon credits are generated from undertaking projects which protect or restore vital habitats such as mangroves, salt marshes, and seagrasses, preventing carbon release and sequestering CO_2. Organizations such as Verra, Gold Standard, and Plan Vivo have developed standards for blue carbon projects and companies can use these credits to offset emissions by purchasing them in voluntary carbon markets. The world's first project to generate blue carbon credits was a community-led mangrove conservation and restoration project in Gazi Bay, Kenya. Proceeds were generated from the sale of credits generated on 117 hectares of mangrove forest.

In 2020, Verra developed the VM0007 methodology to include tidal wetland conservation and restoration activities, which was applied in Cispatá Bay in the Gulf of Morrosquillo, Colombia, resulting in the world's first mangrove conservation project to separately account for soil organic carbon in sediment. The project covers 11,000 ha of mangrove forests and generates approximately 33,000 credits each year, which have been partially purchased by Apple to offset its carbon footprint for 2020 [53].

Macroalgae cultivation, which involves growing seaweeds and kelp or collecting floating algae like sargassum, can be used to increase CO_2 uptake in shallow oceans as it converts dissolved CO_2 into organic carbon through photosynthesis [41]. Harvested macroalgae could be used to create biochar or other durable goods such as bioplastics. The macroalgal biomass could also be harvested, compressed, and sunk to the deep ocean for permanent carbon sequestration. While these are not mainstreamed, three blue carbon offset credit schemes have been implemented in Japan by Yokohama BC Credit, Fukuoka BC Credit and J-Blue Credit for seagrass meadows, macroalgal beds, and macroalgae farming [38].

Ocean fertilization methods aim to increase the production of microalgae in the ocean, which boosts the process where marine organisms convert CO_2 into organic matter. This organic matter sinks and transports carbon into the deep ocean. Adding iron sulfate, nitrogen (urea) or phosphorus can stimulate phytoplankton growth in the ocean. This could be done using vessels to distribute nutrients into nutrient-limited areas of the sea. Artificial upwelling, which involves using pumps or tubes to bring nutrient-rich, cold water from the deep ocean to the surface, is another way to fertilize surface waters [42]. These methods are still in the experimentation stages, and no carbon credits have been issued for these activities.

5.7.5.2 Chemical Approaches

There are two main chemical pathways—Ocean Alkalinity Enhancement (OAE) and Direct Ocean Capture (DOC). OAE methods increase seawater alkalinity to allow for more CO_2 absorption by the surface ocean. This can be done by additions of alkaline materials such as olivine and magnesium hydroxide to seawater or beaches, allowing it to absorb more atmospheric CO_2, which is converted to dissolved bicarbonate and stored away for long periods of time. Isometric issued the world's first independently verified OAE mCDR credits to Planetary Technologies in June 2025. 625 credits were generated from Planetary's Tufts Cove project in Halifax, Nova Scotia and will be delivered to buyers such as Frontier, British Airways, Shopify and Stripe [34].

DOC is another way to remove dissolved CO_2 directly from seawater via electrolysis or electrodialysis, which can be either stored underground or used in various industrial applications. Electrodialysis works by acidifying ocean water to release CO_2 dissolved in the seawater and then capturing it with a membrane.

Brineworks, Captura and Sea02 are some of the companies that are experimenting with different technologies. DOC is potentially more efficient and lower-cost than DAC as the ocean contains 150 times more carbon dioxide than the atmosphere. Captura announced the sale of carbon credits from its first commercial DOC facilities, with Mitsui O. S. K. Lines signing the first offtake agreement for 30,000 tons of mCDRs, before the end of 2030 [7].

The safe deployment of mCDR technologies depends on robust forecasting and monitoring systems capable of supporting accurate carbon accounting. These approaches are influenced by the highly dynamic nature of ocean systems and marine ecosystems, including their inherent uncertainties. Consequently, credible mCDR projects must adopt rigorous frameworks for MRV, alongside equally comprehensive ecological MRV (eMRV) to assess and minimize potential impacts on ocean health and marine biodiversity. As mCDR is an emerging and high-consequence field, the scientific methodologies and operational protocols underpinning these projects are still evolving. A high degree of scientific rigor, transparency, and precaution is essential as new evidence and data become available.

5.8 Chain of Custody Models

A Chain of Custody (CoC) model is a tracking system for products through the entire supply chain, from origin to final product, to ensure that the product characteristics are verifiable. These may include their emission footprint and organic or ethical sourcing status and are extremely important for system integrity and verification of claims.

An 'identity preserved model' is a CoC model in which the materials or products originate from a single source and their characteristics are maintained throughout the supply chain (ISO 2020). One such example is single-origin coffee beans sourced from a specific location. A 'segregated model' is a CoC model in which specified characteristics of a material or product are maintained from the initial input to the final output. Although the material may not be from a single source, addition of material with different characteristics and/or grade is not allowed. For example, a physically segregated biofuel supply chain ties the environmental benefits to the molecules of the biofuel and the environmental attributes of the fuel can only be claimed by the physical user. This approach ensures maximum traceability of the fuel and the accounting procedures to claim the benefits are straightforward. However, physical segregation of supply chains is costly, and dependence on interconnected infrastructure such as pipelines, tanks and fueling systems may be unavoidable in certain cases.

5.8.1 Mass Balance Model

A mass balance model is a CoC model in which materials or products with different sets of characteristics are mixed according to defined criteria. There are partial physical linkages, as certified and non-certified materials can be mixed. Volume-based accounting across defined system boundaries and time periods ensures robust tracking. For example, 20% biofuels can be blended with 80% fossil fuels and environmental credits that are separate from the physical fuel can be credited and traded. This approach allows scalability and lowers the cost as the environmental attributes of the product can be allocated in proportion to the volume of material delivered in the specific batch. The model has high integrity because claims are anchored to real production volumes within the same supply chain.

5.8.2 Book and Claim System

Book and claim system is a CoC model in which the flow of administrative records representing specified characteristics is not necessarily connected to the flow of physical material or product through the supply chain (ISO n.d.). The system functions like energy attribute certificates in the power markets, which decouple the sale of physical electricity from the green certificates.

The system fully decouples the physical molecule of the zero-carbon fuel or material from its environmental attributes such as lower emissions. This allows the sale and purchase of the environmental attributes separately, without buying the physical product, which can be sold to a different buyer. In practice, the producers of a green fuel (e.g., Sustainable Aviation Fuel) may sell the physical fuel to any customer instead of sending it to the buyer. The buyer 'books' the environmental benefit, such as reduced GHG emissions, into a digital registry. The end-user (buyer) can 'claim' the environmental credits from the registry, even if there is no physical connection along the supply chain. These certificate of the environmental attributes can thereafter be used as proof of lowering emissions for sustainability reporting by corporate buyers.

This approach provides the greatest flexibility as there is no physical link and certificates which represent environmental benefits can be traded independently of the underlying product. This allows production and use of low-emission products in locations where they are cheaper to produce without the need for transporting those materials where there is actual demand for these products. As certificates can be traded across sectors and borders, the system allows high flexibility, which is useful to support early-stage market formation. Book and claim models enable scaling up of production even when there is a mismatch between supply and demand in a particular jurisdiction. The book and claim system relies on robust governance, integrity, traceability and transparency to preserve the credibility of the system.

5.8.3 *Functional Examples of Chain of Custody Model*

5.8.3.1 Sustainable Aviation Fuel

Sustainable Aviation Fuel (SAF) is an aviation fuel produced in conformity with sustainability criteria, considering both carbon and environmental factors and is fully fungible with conventional aviation fuel [27]. SAF can reduce CO_2 emissions significantly (80–90% lower) as compared to conventional aviation fuel. It can be produced from hydrotreating waste fats, oils, greases and plant oils (Hydroprocessed esters and fatty acids -HEFA); through the Fischer–Tropsch (FT) process which converts woody biomass or municipal solid waste into syngas and later to liquid fuel; alcohol-to-jet (ATJ) process which includes dehydration and hydrogenation to converts cellulosic biomass or starchy alcohols from corn, and sugarcane into synthetic kerosene, and using renewable electricity to synthesize synthetic hydrocarbons (E-fuels).

The segregated model for CoC for SAF is only used for specific demonstration flights, as it is operationally expensive to segregate the fueling infrastructure. The mass balance model is the most dominant model today for airport-level SAF deployment as it is scalable and compatible with existing fuel infrastructure. Certified SAF is delivered by a refinery to the airport where airlines are refueled with a blend of SAF and conventional aviation fuel. The airline can thereafter claim the corresponding environmental benefits of the contracted SAF volume.

Sustainable Aviation Fuel Certificates (SAFc) represent the environmental attributes of SAF and use a book-and-claim model for decoupling the aviation fuel use from emissions reduction claims. A SAFc is measured in metric tonnes and represents the emissions reductions associated with one metric ton of unblended SAF. Several airports use the mass balance approach for SAF, including London Heathrow airport, Dubai International airport, Singapore Changi airport, Sydney airport and Chicago O'Hare airport. Emirates Airlines, Singapore Airlines and United Airlines regularly claim SAF usage.

SAF is produced at certified facilities using processes such as HEFA, FT and ATJ, and the lifecycle emissions of the SAF produced are calculated using approved methodologies by the International Civil Aviation Organization (ICAO). Independent, third-party certification bodies verify that the accounting methods used comply with suitable certification schemes such as the International Sustainability and Carbon Certification (ISCC) and the Roundtable on Sustainable Biomaterials (RSB). After verification, SAF certificates are issued for the verified SAF volume, and information such as feedstock used, manufacturing pathway, location and vintage is recorded. The SAF is physically blended with conventional aviation fuel (up to a maximum of 50%) for use by an aircraft. The airline physically using the SAF is not credited with the environmental benefits automatically, but the SAFc are sold separately, with the transfer of ownership being recorded in an SAFc registry [47]. An airline (or corporate) can buy the SAFc and, after a claim is made, the certificate is retired.

SAFc are 'unbundled' into two certificates: SAFcA and SAFcE, which represent the same volume of underlying SAF but may only be used to make Scope 1 and 3 emissions reduction claims, respectively. SAF for Air Transport Providers (or SAFcA) represents the Scope 1 emissions claims intended for air transport providers, while SAF End-user Reduction Claims (or SAFcE) represents the end-user (i.e., Scope 3) claims pertaining to one metric ton of unblended SAF which are eligible for voluntary disclosure [47].

SAFc enables airlines and corporates to claim the use of SAF without refueling with SAF, which allows early market scaling as availability of SAF may be geographically restricted to certain airports. This also allows the same fueling infrastructure to be used, making this book-and-claim method flexible as compared to the segregated model. Carbon Offsetting and Reduction Scheme for International Aviation (CORSIA), which is a global, market-based program for the international aviation industry, also allows airlines to reduce their offsetting requirements using SAF under approved sustainability and lifecycle accounting rules as a part of CORSIA Eligible Fuels (CEF), but the book-and-claim system is still under discussion [57].

5.8.3.2 Renewable Natural Gas/Biomethane

Renewable Natural Gas (RNG) or biomethane has the same chemical properties as natural gas but is carbon–neutral. Biomethane is produced from organic material, such as waste food scraps, manure, agricultural, industrial and farm waste through the biological process of anaerobic digestion. Biogas is predominantly methane and after removing impurities like water, carbon dioxide, and sulfur compounds, the methane content increases to over 90 percent, which can be blended with fossil-based natural gas in pipelines.

For RNG, both the mass balance model and book-and-claim model are used. In the mass balance model, physical RNG is mixed with conventional natural gas in the gas network and a certificate is issued to the RNG producer. A company can purchase a volume of mixed gas along with some of these certificates and claim that they consumed an equivalent amount of RNG. In a book-and-claim model, the physical RNG and certificates are delinked and the end-user can 'claim' the environmental benefits of RNG by purchasing the certificates and retiring them. In practice, after a producer injects RNG into the gas network, they are issued with certificates depending on the amount of RNG injected. The end-user can buy only the certificates without purchasing any gas from the network to offset their natural gas consumption purchased from another source. Thus, there is no physical link with the RNG supply chain.

Different systems are used in North America and the European Union. RNG certificates are used in North America (United States and Canada) in the transportation sector to comply with the low-carbon transport fuel mandates such as the Renewable Fuel Standard (RFS) by the U.S. Environmental Protection Agency's (EPA) and the Low Carbon Fuel Standard (LCFS) in California. RNG certificates also allow corporates to offset their corporate emissions and claim the environmental benefits of the use of biomethane in the voluntary markets [3]. In the European Union, the framework is provided by the EU Renewable Energy Directive (RED), which provides a more harmonized system. There are two types of renewable gas certificates: Guarantee of Origin (GOs) and Proof of Sustainability (PoS). Some countries such as Germany, the UK and Austria also have national accounting systems for use of these certificates.

The Guarantee of Origin (GO) for renewable gas is an electronic certificate proving that 1 MWh of biomethane was produced renewably. It provides information such as energy source, location, facility type, production date and time with an aim to ensure transparency. National authorities appoint an issuing body in each country to operate a GO registry which issues GOs to the producer according to the relevant standard when renewable gas is produced and injected into the grid. Once the producer sells the certificate, the GOs are transferred to the registry account of the buyer. As this is a book-and-claim system, certificates can be traded without the physical flow of the renewable gas. The GOs are retired after the final consumer claims the usage of the renewable gas. GOs are primarily used for consumer disclosure by proving the origin of the biomethane and support voluntary and reporting markets.

A Proof of Sustainability (PoS) is a certificate that proves that the biomethane meets the sustainability requirements of the EU RED, such as the source of feedstock for biomethane. The PoS certificate also records the GHG emissions attributed to different parts of the production process and for transportation of biomethane. Unlike GOs, which support voluntary markets and are used for reporting, PoS are used for mandatory compliance markets. If biomethane is used for a compliance market, the PoS must be linked to a GO to avoid double counting. As this is a mass balance model, the PoS certificate must be linked to the trade of physical gas. The European Commission uses the Union Database (UDB) to track and trace the ownership of the PoS [20].

As an example, in Germany, TÜV SÜD and TÜV Nord certify biomethane with GOs using a mass balance approach for grid injection. In the Netherlands, VertiCer issues GOs for biomethane and both mass balance and book-and-claim are used. In the UK, Renewable Gas Guarantees of Origin (RGGOs) use a book-and-claim system which allows corporate users to claim renewable gas without physical delivery.

Certificates of Origin (CoO) is another umbrella scheme used by the European Renewable Gas Registry (ERGaR) that includes both GOs and national biomethane certificates (ERGaR no date b). The scheme facilitates the cross-border transfer of CoOs between participating national biomethane registries and works in a similar way to the GO, with a digital database tracking issuance and ownership. National biomethane registries of Austria, Denmark, Germany, Netherlands, Slovakia, Switzerland and the UK are approved system participants and have access to wider markets.

5.8.3.3 Hydrogen and Its Derivatives

Chain-of-custody (CoC) models are also used for certification of green hydrogen and its derivatives like green ammonia, synthetic kerosene and green methanol.

Renewable Fuel of Non-biological Origin Certificates

Renewable Fuels of Non-Biological Origin (RFNBOs) are synthetic liquid or gaseous fuels, which are created from processes that combine renewable hydrogen and captured carbon. They exclude biomass-derived fuels and must meet strict standards such as use of renewable electricity for their production, additionality, temporal correlation, and geographical correlation with renewable energy used. They must also reduce GHG emissions by at least 70% compared to fossil fuels on a full life-cycle (well-to-wake) basis to be eligible for certification under Renewable Energy Directive III of the EU. RFNBOs play an important role in decarbonization of hard-to-abate sectors such as the maritime and aviation sectors where use of batteries is difficult.

An independent auditor validates the production of RFNBOs as per the approved methodology. The European Commission has approved three certification schemes for RFNBOs: ISCC EU (International Sustainability and Carbon Certification), CertifHy, and REDcert (SPIN 2025). Certificates with information such as the production date and location, feedstock used and GHG intensity of production are issued in units of megajoules (MJ) of energy content (or MWh equivalent) or in units of weight (e.g., kilograms for hydrogen). These certificates are recorded in a national or interoperable EU registry and tracked for transparency. RFNBO certificates are compliance instruments under the EU RED III regulation which are required to qualify for subsidies, meet renewable energy targets, or fulfil fuel mandates. RFNBO certificates are tied directly to the clean fuel molecule being produced and consumed within the regulated system, and hence they cannot substitute for emissions reductions elsewhere.

Hydrogen Guarantees of Origin (H2-GOs)

Hydrogen Guarantees of Origin (H2-GOs) are digital certificates for hydrogen produced from renewable or low-carbon sources which are used for disclosure. Unlike RFNBOs, which have a strict criterion for renewable hydrogen, H2-GOs cover 'fully renewable' and 'low-carbon' hydrogen with different carbon intensity thresholds. H2-GOs are a book-and-claim system and separate the certificate from the physical hydrogen molecule.

Power-To-X (PtX) Certificates

Power-to-X (PtX) is a term used to refer to a suite of technologies which use renewable electricity to produce a wide variety of end products (X) such as green hydrogen, synthetic fuels (e-fuels) and ammonia. PtX certificates prove that end products are produced using renewable energy and the captured CO_2 is sustainably sourced through the production chain. The digital certificate represents that 1 MWh of renewable electricity is used in PtX processes, which guarantees that the final product (e.g., green hydrogen, e-fuels) meets sustainability criteria, which is supported by documentation and verification from independent auditors. Both mass balance and book and claim approaches are used to ensure integrity [32].

5.8.3.4 Low-Carbon Materials Certificates

Bio-Based Feedstock Certificates

Bio-based feedstock certificates verify that a material was produced using bio-based feedstock, such as plant or animal-based raw materials. Bio-based content certification specifies that a portion of fossil-based feedstock (for example, 70%) was replaced with feedstock from renewable biological sources.

ISCC PLUS (International Sustainability and Carbon Certification) is a voluntary certification scheme designed to validate sustainability characteristics of alternative feedstocks. It uses the mass balance model linking physical flows with certificates and allows physical mixing of sustainable (bio/circular) and conventional materials [33]. It does not allow the book-and-claim model for its core certifications. RSB, which is also a certification scheme, allows both a mass balance model and a book-and-claim model for bio-based materials. These certificates are used across industries such as food, chemicals, and packaging for complying with regulatory frameworks, informing customers and disclosures on sustainable sourcing and making sustainability claims. Examples include bio-plastic bottles, which are made from Polylactic Acid (PLA) with 100% bio-based content using corn starch as feedstock, which is certified by RSB. Bio-based lubricants, which use vegetable oils as feedstock and are certified by ISCC, are also used in the industry. Bio-based chemicals used in cosmetics, which are derived from bio-ethanol and are certified by ISCC PLUS, use a USDA BioPreferred label for their final product.

Plastics

Mechanically recycled plastics certification verifies the amount and traceability of recycled content in products. It uses the physical segregation model and mass balance model. Relevant certification standards include EuCertPlast, the Association of Plastic Recyclers (APR) post-consumer recycled content standard, RecyClass and UL 2809 (recycled content verification). These certifications use audits to track recycled plastic along the supply chain, starting from the waste source to the final product, to confirm the recycled material percentages in the final product. These help in meeting regulatory compliance and in claiming lower plastic taxes in countries such as Spain and also allow enhanced access in markets with strict recycled content requirements. Certified circular polymers use ISCC PLUS to verify the recycled content in the product via a mass balance approach. This ensures traceability for materials and verifies that plastic waste is converted back to feedstock.

Steel

CoC models in steel include physical segregation, mass-balance approach and book-and-claim model. ResponsibleSteel provides a certification for steel which includes physical traceability with full segregation for the entire production chain. The CARES Sustainable Constructional Steel (SCS) scheme provides full product traceability extending from manufacturer to end-user, which provides an identity-preserved model for specific projects. The World Steel Association also provides guidelines on using mass balance and book and claim models via transferable 'GHG reduction' certificates. Companies like Tata Steel Nederland offer specific products such as *Zeremis Carbon Lite* based on a mass balance model which allows sale of environmental attributes. JSW Steel offers a certificate-based low-carbon steel product called *GreenEdge Steel,* which is based on a book and claim approach [37]. ArcelorMittal's *XCarb Green Steel Certificates*, which are audited by DNV, also allow customers to claim the environmental benefits [5]. Consumers of steel can purchase and use these certificates to claim Scope 3 emission reductions for their corporate targets.

Aluminum

Low-carbon aluminum production can be certified through ASI (Aluminium Stewardship Initiative) CoC Standard, which uses a mass balance approach. The CoC Standard applies through the value chain (from mining or recycling through to final product). Companies also use their own labels for low-carbon aluminum such as 'Reduxa' by Hydro and 'RenewAl' by Rio Tinto, which are certified by third parties.

5.9 Accelerating Climate Action Through Markets

Strengthening carbon pricing mechanisms and a stronger price signal is essential for the success of market-based solutions. Expanding carbon taxes in jurisdictions where it has not been implemented, gradually increasing carbon taxes, expanding sector coverage and introducing carbon trading schemes in a phased manner while addressing concerns on social equity and competitiveness will help in accelerating climate action. Establishing standardized criteria that address transparency, permanence, and additionality in voluntary carbon markets will ensure a steady stream of high-quality carbon credits that create real, verifiable climate impact. Harmonized measurement standards with robust mechanisms and methodologies for establishing baselines with third-party verification and interoperability of certification schemes will enhance trust in market mechanisms for voluntary markets. Enhanced enforcement of existing rules, enabling transparency through use of blockchain-based digital MRV systems, introducing tokenized systems, lowering transaction costs and improved efficiency will lead to expansion of markets through building trust in the mechanism.

International coordination on carbon markets has commenced through operationalizing the Article 6 framework, which is estimated to accelerate mitigation projects through enabling functional carbon markets. Building global digital infrastructure and registry systems for reporting, tracking and verification of environmental credit data globally needs to be fast-tracked. Capacity building for adoption of market mechanisms in developing countries through technical assistance, developing human resources, skill enhancement, and support for building institutional frameworks will enable development of high-integrity markets which will attract private investment. In developing countries, market mechanisms need to be integrated with broader economic transformation goals while addressing development challenges. Public procurement mandates for low-emission products create demand which incentivizes private investment in low-emission production technologies. Long-term contracts and clear policy signals for setting up tradable markets provide a multi-year view and provide investment certainty. Business models also need to evolve from offsetting to insetting to promote decarbonization of the entire value chain and enhance the adoption of market-based approaches for climate action.

5.10 Conclusion

Market mechanisms are not a panacea for accelerating climate action. Instead, it is only one tool in the range of options available to policymakers. As there is a very low price on carbon in many countries, financial incentives for preventing CO_2 emissions are insufficient. There are gaps in sectoral coverage and geographies. Quality and integrity of carbon credits and the risk of greenwashing are major concerns. Challenges include MRV and ensuring that offsets represent long-lasting emission reduction. Environmental benefits from projects have been found to be overstated by several times in underregulated markets. Scaling challenges include liquidity constraints for high-quality credits. Lack of standardization, poor integrity and transparency also create challenges for accelerating market-based solutions.

Despite these challenges, markets are indispensable for accelerating climate action as it provides economic incentives for investment and innovation, leading to scaling up climate solutions across sectors. Compliance carbon markets, voluntary carbon markets and Article 6 mechanisms serve distinct stakeholders in different jurisdictions and act in a complementary manner. These must be supported by robust MRV and high transparency for delivering real climate benefits. Markets also create incentives for renewable energy, energy efficiency, circular economy practices, biodiversity conservation and environmental services, thereby delivering enhanced value across sectors. Nature-based market mechanisms support and deliver healthy ecosystems which can provide climate, ecological, and socio-economic benefits. CDR markets will increasingly become important as they provide net-negative emissions. Chain of Custody models and product traceability systems also facilitate ease of procurement of low-carbon products. High environmental integrity is important for the long-term success of market mechanisms to deliver climate solutions. Going forward, market mechanisms must be expanded across sectors and countries as they deliver credible results which support positive climate outcomes and advance the objectives of the Paris Agreement.

References

1. Act4Water (n.d.) Act4water Standard: Summary. https://www.act4water.org/wp-content/uploads/2025/02/Act4water-standard_summary-v3-ENG.pdf. Accessed 10 June 2026
2. Act4Water (n.d.) CAPs (Credit Aqua Positive) Water Credits. https://www.act4water.org/en/water-credits/. Accessed 10 June 2026
3. AFS Energy (n.d.) North American Biomethane Credits: Navigating RINs, LCFS, and State-Level Incentives. https://www.afsenergy.nl/blog-post/north-american-biomethane-credits-navigating-rins-lcfs-and-state-level-incentives. Accessed 18 Sept 2025
4. Anaxee Digital Runners (2025) How Biochar Carbon Credits Work: From Production to Certification. https://anaxee.com/biochar-carbon-credits/. Accessed 16 Sept 2025
5. ArcelorMittal (n.d.) XCarb Green Steel Certificates. https://projects.arcelormittal.com/media/i1xm0zsk/xcarb-green-steel-certificates-2110.pdf. Accessed 18 Sept 2025
6. Barbose GL (2025) Berkeley Lab publishes status update on state clean electricity standards. Lawrence Berkeley National Laboratory, Energy Markets and Planning. https://emp.lbl.gov/news/berkeley-lab-publishes-status-update-state-clean-electricity-standards. Accessed 27 March 2026
7. Captura (2025) Captura Announces Sale of Carbon Removal Credits and Strategic Partnership With Mitsui O.S.K. Lines. https://capturacorp.com/sale-of-carbon-credits-and-partnership-with-mol/. Accessed 20 June 2026
8. CarbonCredits.com (2025) The Biochar Gold Rush: Why Companies are Scrambling to Lock in Carbon Credits. https://carboncredits.com/the-biochar-gold-rush-why-companies-are-scrambling-to-lock-in-carbon-credits/. Accessed 20 June 2026
9. CCARDESA (n.d.) Agroforestry: Silvo-Pasture. https://www.ccardesa.org/agroforestry-silvo-pasture. Accessed 22 Sept 2025
10. Circular Action Hub (n.d.) Circular Credits Mechanism. https://www.circularactionhub.org/certification/circular-credits-mechanism/. Accessed 22 Sept 2025
11. Clean Energy Regulator (n.d.) Renewable Energy Target. Australian Government. https://cer.gov.au/schemes/renewable-energy-target. Accessed 18 Sept 2025
12. Climate Action Reserve (n.d.) Grassland Protocol. https://www.climateactionreserve.org/how/protocols/ncs/grassland/. Accessed 23 Sept 2025
13. Earth.Org (n.d.) Soil Carbon Credits: Promises and Problems. https://earth.org/soil-carbon-credits-the-promises-and-uncertainties-of-a-new-climate-market/. Accessed 23 Sept 2025

14. Eco-Markets Australia (2024) New Grazing Land Management Methodology Launched Under Reef Credit Scheme. https://eco-markets.org.au/2024/09/17/new-grazing-land-management-methodology-launched-under-reef-credit-scheme/. Accessed 22 Sept 2025
15. Eco-Markets Australia (n.d.) Reef Credits. https://eco-markets.org.au/reef-credits/. Accessed 22 Sept 2025
16. Electric Power Research Institute (n.d.) Ohio River Basin Trading Project: Overview. https://wqt.epri.com/overview.html. Accessed 26 Sept 2025
17. Environmental Defense Fund and Carbon Limits (n.d.) Carbon Credits for Livestock Methane Reductions. https://library.edf.org/AssetLink/1o85xwykirw3r30r7p440dgsk812c0g7.pdf. Accessed 16 June 2026
18. Envirotech Online (n.d.) The First List of the 57 Water Quality Trading Programs Worldwide Is Being Released by the World Resources Institute. https://www.envirotech-online.com/news/water-wastewater/9/ege-elektronik/the-first-list-of-the-57-water-quality-trading-programs-worldwide-is-being-released-by-the-world-resources-institute/5731. Accessed 12 Sept 2025
19. European Renewable Gas Registry (n.d.) Renewable Gas Certificates. https://renewablegascertificates.org/renewable-gas-certificates/. Accessed 19 Sept 2025
20. European Renewable Gas Registry (n.d.) Renewable Gas Certification. https://www.ergar.org/ergar-schemes/ergar-coo-scheme/. Accessed 19 Sept 2025
21. Fastmarkets (2026) Article 6, CORSIA Markets Will Be Shaped by Accounting Preferences. https://www.fastmarkets.com/insights/article-6-corsia-markets-will-be-shaped-by-accounting-preferences/. Accessed 26 June 2026
22. Fortune Business Insights (n.d.) Voluntary Carbon Credit Market. https://www.fortunebusinessinsights.com/voluntary-carbon-credit-market-115896. Accessed 21 Sept 2025
23. Google (2020) 24/7 by 2030: Realizing a Carbon-free Future. https://sustainability.google/reports/247-carbon-free-energy/. Accessed 21 Sept 2025
24. ICAP (2026) Emissions Trading Worldwide: Status Report 2026. International Carbon Action Partnership, Berlin. https://icapcarbonaction.com/en/publications/emissions-trading-worldwide-icap-status-report-2026. Accessed 21 June 2026
25. InPlanet (2026) World's First Enhanced Rock Weathering Carbon Removal Credits Issued. https://inplanet.earth/press/worlds-first-enhanced-rock-weathering-carbon-removal-credits-issued/. Accessed 26 June 2026
26. Interesting Engineering (2025) 6.75 Million Tons: Microsoft Signs World's Largest CO_2 Removal Deal. https://interestingengineering.com/innovation/microsoft-largest-carbon-removal-deal. Accessed 10 June 2026
27. International Air Transport Association (2024) SAF Handbook. https://www.iata.org/contentassets/d13875e9ed784f75bac90f000760e998/saf-handbook.pdf. Accessed 21 Sept 2025
28. International Institute for Environment and Development (n.d.) Biocredit catalogue—A collection of biocredit developers and schemes. https://www.iied.org/sites/default/files/pdfs/2023-12/22201iied.pdf. Accessed 21 Sept 2025
29. International Organization for Standardization (2020) ISO 22095:2020 Chain of Custody—General Terminology and Models. https://www.iso.org/obp/ui/#iso:std:iso:22095:dis:ed-1:v1:en. Accessed 27 March 2025
30. International Organization for Standardization (n.d.) ISO/DIS 13659 Chain of Custody—Book and Claim Model. Draft International Standard. https://www.iso.org/obp/ui/en/#iso:std:iso:13659:dis:ed-1:v1:en. Accessed 27 March 2025
31. International Peatland Society (n.d.) What Are Peatlands? https://peatlands.org/peatlands/what-are-peatlands/. Accessed 23 Sept 2025
32. International PtX Hub (2023) Certification for Green Hydrogen and PtX. Briefing. https://ptx-hub.org/wp-content/uploads/2023/05/International-PtX-Hub_202305_Certification-for-green-hydrogen-and-PtX.pdf. Accessed 22 Sept 2025
33. International Sustainability and Carbon Certification (2025) ISCC EU Mass Balance Guidance Document. Version 1.0. https://www.iscc-system.org/wp-content/uploads/2025/04/ISCC-EU-Mass-Balance-Guidance-Document_Version-1.0.pdf. Accessed 28 Sept 2025

34. Isometric (2025) World's First Verified Ocean Alkalinity Enhancement Credits. https://isometric.com/writing-articles/worlds-first-verified-ocean-alkalinity-enhancement-credits. Accessed 27 Sept 2025
35. Isometric (n.d.) Direct Air Capture. Protocol. https://registry.isometric.com/protocol/direct-air-capture. Accessed 16 June 2026
36. IUCN UK Peatland Programme (n.d.) Peatland Code. https://www.iucn-uk-peatlandprogramme.org/peatland-code. Accessed 23 Sept 2025
37. JSW Steel (n.d.) GreenEdge. https://www.jswsteel.in/sustainability/greenedge. Accessed 26 June 2026
38. Kuwae T, Watanabe A, Yoshihara S, Suehiro F, Sugimura Y (2022) Implementation of blue carbon offset crediting for seagrass meadows, macroalgal beds, and macroalgae farming in Japan. Mar Policy 138:104996. https://www.sciencedirect.com/science/article/pii/S0308597X22000434. Accessed 30 Sept 2025
39. MSCI (2026) Carbon Credits Come of Age in 2025. MSCI Research and Insights. https://www.msci.com/research-and-insights/blog-post/carbon-credits-come-of-age-in-2025. Accessed 26 June 2026
40. NSW Department of Climate Change, Energy, the Environment and Water (n.d.) Biodiversity Stewardship Agreements. Biodiversity Offsets Scheme. https://www.environment.nsw.gov.au/topics/animals-and-plants/biodiversity-offsets-scheme/about/how-scheme-works/biodiversity-stewardship-agreements. Accessed 19 Sept 2025
41. Ocean Visions (n.d.) Macroalgae Cultivation and Carbon Sequestration. https://oceanvisions.org/macroalgal-cultivation/. Accessed 30 Sept 2025
42. Ocean Visions (n.d.) Ocean-Based Carbon Dioxide Removal: Microalgae Cultivation. https://oceanvisions.org/microalgae-cultivation/. Accessed 30 Sept 2025
43. PCX Solutions (n.d.) Plastic Crediting Process. https://www.pcxsolutions.org/plastic-crediting-process. Accessed 23 Sept 2025
44. Puro.earth (2025) Enhanced Rock Weathering: Methodology for CO2 Removal. Edition 2025 v.1, draft. https://7518557.fs1.hubspotusercontent-na1.net/hubfs/7518557/ERW%20Standards/ERW%202025/ERW%20Edition%202025%20v1.pdf. Accessed 26 June 2026
45. Regreener (2026) The 5 Best Direct Air Capture Carbon Credit Projects of 2026. https://www.regreener.earth/blog/the-5-best-direct-air-capture-carbon-credit-projects-of-2026. Accessed 16 June 2026
46. RMS Certified (n.d.) ARC Certificates. Recycled Material Standard, GreenBlue. https://www.rmscertified.com/about/credit/. Accessed 27 Sept 2025
47. SAFc Registry (n.d.) VI. SAF Certificates. User Guide. https://docs.safcregistry.org/user-guide/vi.-saf-certificates. Accessed 23 Sept 2025
48. Senken (n.d.) The Complete Guide to REDD+ Carbon Credits and Forest Protection. https://www.senken.io/glossary/redd-carbon-credits. Accessed 27 March 2026
49. Senken (n.d.) Complete Guide to Direct Air Capture Carbon Removal. https://www.senken.io/glossary/direct-air-capture-dac. Accessed 10 June 2026
50. SORT Consultancy (n.d.) EPR Certificates and Credit Trading 2025: India's Circular Economy. https://sortconsultancy.com/blogs/epr-certificates-credit-trading-guide-2025. Accessed 21 June 2026
51. Stavins, R.N., Chan, G., Stowe, R. and Sweeney, R. (2012) The U.S. Sulphur Dioxide Cap and Trade Programme and Lessons for Climate Policy. Vox. https://www.hks.harvard.edu/publications/us-sulphur-dioxide-cap-and-trade-programme-and-lessons-climate-policy. Accessed 19 Sept 2025
52. Swiss Power-to-X Collaborative Innovation Network (2025) EU Commission Approves RFNBO Certification Schemes. https://power-to-x.swiss/2025/02/11/eu-commission-approves-rfnbo-certification-schemes/. Accessed 27 March 2026
53. The Tokyo Foundation (n.d.) Ensuring the Sustainability of Blue Carbon Ecosystems. https://www.tokyofoundation.org/research/detail.php?id=859. Accessed 30 Sept 2025
54. Thompson, P. (2025) An Introduction to Jurisdictional REDD+. Sylvera. https://www.sylvera.com/blog/an-introduction-to-jurisdictional-redd. Accessed 27 March 2026

55. UNEP Copenhagen Climate Centre (n.d.) Article 6 Pipeline: Global Data. https://article6pipeline.unepccc.org/global-data. Accessed 15 June 2026
56. UNFCCC (n.d.) International Registry and Additional Registry Services (ARS). https://unfccc.int/process-and-meetings/the-paris-agreement/article-6/article-62/international-registry-and-ars. Accessed 23 Sept 2025
57. Uniting Aviation (n.d.) Book and Claim Explained: Sustainable Aviation Fuels Accounting and CORSIA. International Civil Aviation Organization. https://unitingaviation.com/news/environment/book-and-claim-explained-sustainable-aviation-fuels-accounting-and-corsia/. Accessed 27 March 2026
58. US Environmental Protection Agency (n.d.) Water Quality Trading. https://www.epa.gov/npdes/water-quality-trading. Accessed 10 Sept 2025
59. World Bank (2018) Carbon Accounting Tools for Sustainable Land Management. https://documents1.worldbank.org/curated/en/318251544164909341/pdf/132767-6-12-2018-14-1-54-SLMFullReportFINAL.pdf. Accessed 27 Sept 2025
60. World Bank (2024) Plastic Credits at a Glance: Product Overview Series for Financial Instruments. https://thedocs.worldbank.org/en/doc/411ebaec936068e4bb62a0e40ebce522-0320072024/original/Product-Overview-Plastic-Credits-FINAL.pdf. Accessed 27 September 2025
61. WWF (n.d.) Piloting Wildlife Credits in Tanzania. https://wwfint.awsassets.panda.org/downloads/wwf_factsheet_tanzania.pdf. Accessed 27 March 2026
62. WWF (n.d.) Biodiversity Certificates to Conserve French Forests. https://wwfint.awsassets.panda.org/downloads/wwf_factsheet_french-forests.pdf. Accessed 27 March 2026

Chapter 6
Technological Solutions for Climate Change

Abstract This chapter spotlights technological solutions for climate change. Electrification, grids, energy storage, demand flexibility, digitalization and artificial intelligence technologies are emerging as effective climate solutions. Low-emission technologies for power generation, batteries and chargers are presented. Emerging road transportation, aviation and shipping technologies are highlighted using the latest global examples. Industrial decarbonization technologies for low-emission steel, cement, chemicals and textiles are discussed. Technologies for lowering GHG emissions from agriculture, livestock farming, nature-based solutions and solar radiation modification are also presented. The chapter concludes that technological solutions are advancing at a fast pace and need to be scaled up rapidly.

Keywords Energy storage · Advanced solar cell · Small modular reactors · Batteries · Industrial decarbonization · Solar radiation modification

6.1 Introduction

Technology is a critical enabler and an essential part of climate mitigation and adaptation solutions. Climate technologies have a dual role: climate mitigation and building climate resilience. Renewable energy (RE) technologies replace fossil fuel-based energy sources and mitigate CO_2 emissions. Energy efficiency technologies lower energy consumption and energy bills of the organization and households. Widespread adoption of low-emission and near-zero-emission technologies reduces environmental pollution, thereby impacting air and water quality. Monitoring technologies such as remote sensing, satellite-based imaging, environmental and GHG monitoring systems allow tracking of CO_2 emissions. Technologies such as early warning systems for extreme weather, water management, urban design, and adaptive farming are critical for climate adaptation. Different climate technologies are applicable for different sectors and a mix of nature-based climate solutions and manmade technologies is required for addressing climate change. A portfolio approach that

K. Narula, *The Climate Catalyst*, Advances in Sustainability Science and Technology,
https://doi.org/10.1007/978-981-92-5213-8_6

implements the most appropriate technology in the sector, depending on their suitability, availability and cost, can advance climate action. Along with technological innovation, supporting policies and financing options are essential for scaling up low-emission technologies.

This chapter discusses technological solutions for climate change. Section 6.2 identifies the emerging technological shifts, including electrification, grids, energy storage, demand flexibility, digitalization and artificial intelligence. Section 6.3 focuses on low-emission technologies for power generation, including next-generation solar, offshore floating wind, and advanced nuclear reactors, using examples. Section 6.4 deep dives into next-generation batteries and charging infrastructure. Section 6.5 presents emerging road transportation, aviation and shipping technologies supported by global examples. Section 6.6 focuses on industrial decarbonization technologies for low-emission iron and steel, cement and concrete, chemicals and plastics and textiles. Agriculture is a large source of GHG emissions and technologies such as feed additives for cattle, plant engineering, precision fermentation and carbon-fixing microorganisms are presented in Sect. 6.7. Section 6.8 discusses technologies for implementing nature-based solutions and Sect. 6.9 presents the futuristic technology of solar radiation modification before concluding the chapter.

6.2 Emerging Technological Shifts

6.2.1 System of Systems Approach

A system of systems approach to climate action acknowledges that there is a need to recognize the interdependencies between energy, water, food, transport, urbanization and infrastructure for optimal climate solutions. Solving one problem through technology scale-up should not lead to another problem such as excessive extraction of natural resources, generation of non-biodegradable waste and degradation of the environment. An integrated approach hinges on interoperability of systems, common data standards and protocols, sector coupling of power, heat, cooling, gas, hydrogen and fuel networks, decentralization of energy systems, implementing circular feedback loops and building resilience through redundancy.

6.2.2 Electrification of Sectors and Grids

An emerging technological theme is electrification of energy demand through fuel switching in different sectors. This requires flexible grids with additional grid capacity. Grids also need to become smarter, resilient and intelligent, allowing for bidirectional flow of electricity. Long-distance transmission of electricity using high voltage direct current (HVDC) links and intercontinental connectors, which could include subsea interconnectors, is required for long-distance transmission of power. Grid-enhancing technologies, including dynamic line reading which uses Internet of Things (IoT) sensors to monitor real-time weather conditions to allow higher electricity transmission using the same cables, development of advanced conductors which have higher capacity as compared to steel cables, are emerging. Software-defined grids and virtual power plants, which aggregate distributed energy resources, use of predictive maintenance and self-healing grids are essential components of advanced future grids.

6.2.3 Energy Storage and Demand Flexibility

Following the increased demand for electricity, there is an emerging need for technologies for energy storage and electricity demand flexibility. Use of batteries for electricity storage has expanded beyond short-term smoothing of fluctuations and backup power for a few hours into long-duration energy storage. Behind-the-meter energy storage is also becoming an active source of electricity, which is transforming passive consumers into active 'prosumers' utilizing the bidirectional flow capability of grids. AI-driven demand flexibility in households, industries and data centers allows for quick demand-side responses through load flexibility, easing the burden on power systems.

6.2.4 Electrons and Molecules

The synergy between electrons (electricity) and molecules (low-emission fuels) is vital in the energy transition phase. Green electrons can be used directly to electrify heating and cooling, replacing the use of fossil fuels. However, for some hard-to-abate sectors such as steel, aviation and shipping, there is a role for hydrogen, ammonia and E-fuels as the sector needs high-energy-density energy sources. Electrolyzers allow conversion of electricity generated from renewable energy (RE) into hydrogen, which can be stored and transported as ammonia. These green molecules can then be used to produce synthetic fuels such as sustainable aviation fuels. Natural gas fired furnaces can be replaced by electric arc furnaces and plasma torches for high-temperature requirements. Use of electricity in the industry also allows chemical reactions at lower temperatures, which reduces energy wastage and fuel use.

6.2.5 Digitalization and Artificial Intelligence

Digitalization and Artificial Intelligence (AI) based systems are emerging as the backbone for monitoring the climate transition. Digitization enables an automated monitoring system which brings increased transparency and creates accountability. High-fidelity observation systems with multimodal sensing through combining high-resolution satellite data with on-ground IoT sensors allow high measurement accuracy, near-real-time methane leak detection and remote monitoring of forest fires, giving a lead time for response. Digitalization has enabled tokenization, while distributed ledgers allow tracking of carbon emission certificates, increasing the credibility of digital trading. Smart contracts permit immutable ledgers and blockchains can be used for financialization of environmental attributes. AI-powered digital twins allow simulation of different climate scenarios for climate risk assessment and enable pricing of physical climate risks into insurance premiums. This enables development of resilient infrastructure and addresses climate adaptation.

6.3 Low-Emission Power

RE technologies are fast replacing fossil fuels for electricity generation. Table 6.1 shows the comparative assessment of different technologies for power generation with their advantages, limitations and future outlook. Table 6.2 compares different technologies for low-emission power generation in terms of technical and cost parameters.

As shown in Table 6.2, onshore wind is the cheapest source of electricity globally, followed by solar PV. There has been a long-term cost decline for solar PV, and its Levelized cost of electricity (LCOE) is almost 90% lower than in 2010. While there is a short-term trend of increasing LCOE for solar PV and onshore and offshore wind in 2023–2024, further cost reductions are expected by 2030, subject to supply chain risks.

6.3.1 Next-Generation Solar

Solar energy is the fastest-growing RE source over the past decade. It is modular and scalable. The technology has matured and is suitable for stand-alone and utility-scale applications. Solar PV is being rapidly deployed across the world to generate electricity, with the largest additions happening in China. The two main challenges are low efficiency of solar PV panels and low-capacity factor.

6.3.1.1 Perovskite Solar Cells

Solar panels made from amorphous or crystallized silicon wafers have an efficiency of 15–22% and require energy-intensive manufacturing processes. Advances in technology have led to improvements, and LONGi achieved a 27.81% efficiency for silicon solar cells in 2025 [34]. New Perovskite Solar Cells (PSCs) made of

Table 6.1 Comparison of technologies for low-emission power generation

Technology	Advantages	Limitations	Future outlook
Hydropower	Provides flexible 'firm' power and large-scale energy storage	High ecological impact; high capital expenditure (CapEx); vulnerable to climate-induced droughts	Pumped Storage Hydropower (PSH) can act as a battery for the grid
Solar photovoltaic (PV)	Lowest cost RE source; modular installation; highly scalable from residential rooftop to utility	Intermittency; land use concerns; requires battery backup for 'firm' power	Faster deployment; Higher efficiency cells under development
Concentrated solar power (CSP)	High temperature heat generation with integrated thermal storage allows power generation after sunset	High CapEx; requires direct sunlight and water for cooling	Hybrid plants (PV + CSP) being developed to provide 24/7 power
Onshore wind	Complements solar power; allows simultaneous land use for farming	High noise; not modular; high installation cost	Old turbines gradually being replaced with taller and higher capacity turbines
Offshore wind	Can capture stronger and consistent winds; Larger size and capacity; No land availability trade-offs	High installation and O&M costs due to harsh marine environment; grid connection challenges	Floating offshore wind platforms being deployed globally
Bioenergy	Can be carbon-neutral and carbon negative	Competition with food crops; biodiversity loss	Advanced biofuels from waste/algae under development
Geothermal	24/7 baseload power; small surface footprint	Location-specific; high drilling costs; risk of seismic activity	Enhanced geothermal systems being developed

organic–inorganic hybrid material are emerging as an alternative, with Perovskite (single-junction) solar cells achieving an efficiency of close to 27%.

Perovskite material is a hybrid organic–inorganic lead or tin halide that uses a unique material layer characterized by its specific crystal structure [28]. This structure is defined by the chemical formula ABX3, where:

- A is a large monovalent organic or inorganic cation (like methylammonium, formamidinium, or cesium) located in the corner positions of the cubic structure
- B is a small divalent metal cation (typically lead or tin) situated in the center of the structure
- X is a halide anion (such as iodide, bromide, or chloride) that bonds to both A and B cations.

Table 6.2 Comparison of low-emission power generation technologies (2010 vs. 2024)

Technology	Total installed cost (2024 US$/kW)			Capacity factor (%)			LCOE (2024 US$/kWh)		
	2010	2024	% change (%)	2010	2024	% change (%)	2010	2024	% change (%)
Hydropower	1,494	2,267	52	44	48	9	0.044	0.057	30
Solar PV	5,283	691	−87	15	17	13	0.417	0.043	−90
CSP	10,703	3,677	−66	30	41	37	0.402	0.092	−77
Onshore wind	2,324	1,041	−55	27	34	26	0.113	0.034	−70
Offshore wind	5,518	2,852	−48	38	42	11	0.208	0.079	−62
Bioenergy	3,082	3,242	5	72	73	1	0.086	0.087	1
Geothermal	3,083	4,015	30	87	88	1	0.055	0.060	9

Data source IRENA [40]

The material has optoelectronic properties and is a strong light absorber with high absorption coefficients, which allow the cells to be very thin, flexible, and lightweight. These cells also have lower manufacturing costs and easier manufacturing processes using solution-processing techniques such as spin-coating and vacuum deposition. They are highly tolerant of defects and impurities, can be printed at room temperature, and coated onto flexible substrates like plastics. However, cells degrade rapidly when exposed to the atmosphere and have a much shorter life span of about 2–5 years, which is about 4–5 time lesser than silicon-based cells.

Perovskite-silicon tandem cells are being developed that stack a top layer of perovskite material onto a conventional silicon bottom layer. This cell absorbs different parts of the solar spectrum: perovskite absorbs visible light while silicon absorbs infrared light and can achieve higher conversion efficiency. In April 2025, LONGi reported 34.85% efficiency in perovskite-silicon tandem cells, surpassing single-junction silicon cells [44].

Perovskite solar cells are a breakthrough in solar PV technology due to high efficiency and low-cost, flexible manufacturing. Stability and the life of these cells are being addressed, and once these issues are resolved, perovskite tandem cells could significantly accelerate the global shift to cleaner, cheaper solar power.

6.3.1.2 Floating Solar PV

A major limitation of ground-mounted solar PV farms is that they have a large surface footprint and compete for agricultural land with or urban ecosystems. Floating solar farms overcome this limitation and have grown rapidly in the past few years. As the bottom part of the solar panels is exposed to water, the cooling effect improves panel performance (versus land-based systems), leading to higher efficiency. Covering water bodies such as dam reservoirs with floating solar PV panels can also reduce water evaporation significantly, resulting in increased hydroelectricity generation.

Floating solar farm installations are projected to rapidly grow in the Asia Pacific region. In late 2024, state-owned China Energy Investment Corporation commissioned a 1GW floating solar PV project off Dongying City in Shandong Province, which was claimed to be the world's largest floating solar farm. Located 8 km offshore near the Bohai Sea, the project spans 1,223 hectares, uses nearly 3,000 PV platforms and a 66 kV offshore–onshore cable [60]. Other projects include the 600 MW Omkareshwar project (phased development) situated on the Narmada river's reservoir in India and the 192 MW Cirata floating solar PV project on a reservoir in West Java province in Indonesia.

6.3.1.3 Space-Based Solar Power

Space-based solar power (SBSP) is a futuristic option to provide 24 $\times$ 7 clean power. Satellites in geostationary orbits have access to uninterrupted as well as higher intensity of sunlight as compared to the Earth's surface. Solar panels on satellites in geostationary orbit can convert energy into microwaves or lasers, and power can be beamed back to Earth-based receivers wirelessly. In 2023, the California Institute of Technology successfully demonstrated wireless transmission of power from space to Earth through its Microwave Array for Power-transfer Low-orbit experiment, opening the doors for a successful commercial venture [52]. Similar projects are being planned by China, Europe, the UK and the US in the next few years.

6.3.2 Offshore Floating Wind Turbines

Floating offshore wind turbines have a distinct advantage over fixed-bottom offshore wind turbines. Fixed-bottom offshore wind farms are typically restricted to less than 60 m water depth, which is found in the continental shelf, but floating offshore wind farms can be deployed in much deeper waters as they are not installed on the sea floor. They also have access to stronger and more consistent wind resources as they are located in deep ocean, leading to capacity factors exceeding 50%, compared to 30–40% for onshore wind turbines. Floating wind turbines use buoyancy via semi-submersible, spar, or tension-leg platforms, anchors and mooring lines, which have a minimal seabed footprint, unlike fixed-bottom offshore platforms, which have a significant environmental impact due to piling or foundation work. Floating offshore wind farms also have a larger resource potential, as shallow continental shelves, which have high wind profiles, are limited. Lastly, as offshore floating wind farms are out of sight and have less visibility from the shore, that leads to lower opposition to these wind farms.

The installed operational capacity of global offshore wind at the end of 2024 was 75–80 GW, with floating offshore wind capacity being only 250–300 MW. The future pipeline is strong, with over 1,200 GW of planned projects, of which about 300 GW have already been consented, and the remaining 900 GW are under planning. Large commercial-scale floating projects of over 500 MW are expected to be delivered in this decade. The Hainan offshore floating wind farm (average depth 100 m), located 22 km off the coast of Wanning City, Hainan, China, is reported to be the largest planned offshore floating wind farm. It has a planned capacity of 1,000 MW. The first phase of 200 MW is under construction, and the second phase having 800 MW, will be delivered in 2027 [25].

The current LCOE of offshore floating wind is around USD 120–200/MWh as compared to USD 45–75/MWh for fixed-bottom offshore wind. It is expected that these costs will decline rapidly to USD 60–90/MWh by 2035 due to advancements in technology and development of higher capacity wind turbines [39]. Table 6.3 lists the largest operational offshore floating wind farms globally, as of the end of 2025, which shows their growing deployment.

6.3.3 Advanced Nuclear Reactors

Nuclear energy is reemerging as one of the important low-emission electricity technologies and is finding growing policy support among countries. Its key advantages include near-zero life cycle GHG emissions, firm and dispatchable power, smaller land footprint and high energy density, which makes it an ideal energy source for complementing electricity from variable RE sources. The ability for heat and power cogeneration, over 90% capacity factor, and higher fuel utilization rates makes nuclear energy an important source for meeting the goal of global net zero emissions.

6.3.3.1 Small Modular Reactors

Small Modular Reactors (SMRs) with an output of 300 MWe equivalent or lower have reduced fuel requirements as they use Generation II and III Light Water Reactor (LWR) designs. SMRs mostly use low-enriched uranium, but certain advanced designs use higher-enriched fuels, including High-Assay Low-Enriched Uranium (HALEU). The fuel cycle is mostly once-through, and their waste production per unit output is similar to current reactors.

There are several families of SMRs. Light water reactor (LWR) uses ordinary water, both as the coolant and the neutron moderator, while heavy water reactor (HWR) uses heavy water (D_2O), which absorbs fewer neutrons than regular light water. Hence, HWR reactors can efficiently use natural uranium as fuel rather than requiring enriched uranium. High-temperature, gas-cooled reactors (HTGR) type SMRs provide high-temperature heat ($\geq$750 °C) and can be used for electricity

Table 6.3 Largest operational offshore floating wind farms

Project name, country	Commissioned	Capacity (MW)	Notes/parameters	Data source
Hywind Tampen, Norway, about 140 km off the coast	Aug 2023	95	11 × 8.6 MW turbines Water depth 260–300 m Uses TetraSpar floating offshore wind turbine foundation Supplies electricity to offshore oil platforms	[16]
Kincardine offshore wind farm, UK (Scotland), 15 km off the coast of Aberdeen	Mid 2022	50	5 × 9.5 MW + 1 × 2 MW turbines Uses steel semi-submersible floating foundations Electricity transferred to grid	[22]
Hywind Scotland, UK (Scotland), 25 km off the coast of Peterhead	Oct 2017	30	5 × 6 MW turbines Water depth: 95–120 m First commercial floating offshore wind farm	[17]
WindFloat Atlantic, Portugal, located 18 km off the coast of Viana do Castelo	Jul 2020	25	3 × 8.4 MW turbines 100 m water depth First semi-submersible floating offshore wind farm	[56]
Provence Grand Large, France, located 17 km off the coast of Port-Saint-Louis-du-Rhône	Jun 2025	25	3 × 8 MW turbines Uses tension leg platforms Grid connected	[12]
OceanX dual-turbine floating wind platform, China, Qingzhou IV, Yangjiang, Guangdong Province, 70 km from shore	Dec 2024	16.6	2 × 8.3 MW turbines 45 m water depth Dual-rotor 'V' shaped platform with counter-rotating blades design Largest single-capacity floating wind power platform	[4]

generation, industrial applications and cogeneration. 68 SMRs, of which 22 are water-cooled and 46 are non-water-cooled, were under development in 2024 [35].

SMRs have factory fabrication and short construction times as they are assembled on site. Due to their modular design, the capacity can be increased to match the designed output and they can be deployed closer to the demand centers. Microreactors (a subset of SMRs), which generate up to 15–20 MWe, are suitable as a backup power supply or to supply electricity in microgrids, remote areas, for disaster recovery, and critical service restoration. Thirteen designs of microreactors are currently under development [35]. SMRs require lower initial investment for installation, which broadens the investor base and is attracting private investment. These are expected to play an important role in decarbonizing hard-to-abate sectors and to provide energy for industrial heat, production of hydrogen, desalination and base-load power for electricity grids.

Since December 2023, China has been operating a 200 MWe High Temperature Gas-Cooled Reactor—Pebble-bed Module (HTR-PM) which has two small reactors (each of 250 MWt) that drive a single 210 MWe steam turbine [83]. It uses helium as coolant and graphite as the moderator. Each reactor is loaded with more than 400,000 spherical fuel pebbles (60 mm in diameter) which contain 7 gms of fuel enriched to 8.5%. The reactor generates high-temperature steam up to 500° and electricity [85].

Russia has been operating RITM-200 SMR, a 175 MWt (55 MWe) Generation III+ integrated pressurized water reactor (PWR) since 2020. It uses up to 20% enriched uranium and has a 40-year service life. The reactor is used on Russia's Project 22220 four nuclear icebreakers (*Arktika, Sibir, Ural, Yakutia*) in pairs. The land-based version with a capacity of 55 MWe, RITM-200N, is currently under construction in Ust-Kuyga, Yakutia, and expected to be commissioned by 2028. RITM-200S reactors, planned to be used as floating power units for a large mining and processing facility in Chukotka, are also being developed with an expected commissioning date of 2029 [66].

6.3.3.2 Generation IV Systems

SMRs use LWR technology while Generation IV systems go beyond conventional water-cooled designs. Generation IV nuclear reactors operate at higher temperatures and use alternative coolants such as helium, sodium, lead, molten salt. Fuel types include HALEU, plutonium, thorium and recycled fuels, which operate in closed or partially closed fuel cycles. Sizes range from 150 to 1500 MWe, with the lead-cooled one available as a replaceable cassette of 50–150 MWe with long core life (15–20 years without refueling). These are currently in the design stage. An international task force—the Generation IV International Forum (GIF) is sharing R&D to develop Generation IV nuclear reactor designs, as shown in Table 6.4 [24].

The outlook for deployment of advanced nuclear technology is strong, with commercial deployment expected to pick up in the second half of this century. In the next decade, SMRs are expected to contribute significantly to new nuclear capacity and advanced reactor designs are likely to be demonstrated with the first commercial units becoming operational. Announced projects include a 50 MW SMR in Oak Ridge, Tennessee to be built by Kairos Power, which would supply electricity to Google's data centers in the U.S. in 2030 [63]. In Sweden, Vattenfall plans to order either five BWRX-300 reactors from GE Vernova or three Rolls-Royce SMRs, to add to the Ringhals plant in southwest Sweden [64].

Table 6.4 Generation IV reactor designs under development

Reactor type, neutron spectrum	Coolant	Temp (°C)	Pressure*	Fuel and fuel cycle	Size (MWe)
Gas-cooled fast reactors, fast	Helium	850	High	U-238+, Closed, on-site	1200
Lead-cooled fast reactors, fast	Lead or Pb-Bi	480–570	Low	U-238+, Closed	20–180**; 300–1200; 600–1000
Molten salt fast reactors, Fast	Fluoride salts	700–800	Low	UF in salt, closed	1000
Molten salt reactor—advanced high-temperature reactors, thermal	Fluoride salts	750–1000	Low	UO_2 particles in prism, open	1000–1500
Sodium-cooled fast reactors, fast	Sodium	500–550	Low	U-238 & mixed oxide (MOX) fuel, closed	50–150; 600–1500
Supercritical water-cooled reactors, thermal or fast	Water	510–625	Very high	UO_2 open (thermal); closed (fast)	300–700; 1000–1500
Very high temperature gas reactors, thermal	Helium	900–1000	High	UO_2 prism or pebbles, open	250–300

Notes * High pressure is 7–15 MPa; + Implies with some U-235 or Pu-239; ** Model with long cassette core life of 15–20 years or replaceable reactor module
Data source World Nuclear Association [84]

6.3.4 Other Low-Carbon Energy Sources

Beyond solar, wind and hydro power, other low-carbon energy sources include modern bioenergy, geothermal, ocean energy and waste-to-energy, which could play a role in a net zero world. A summary of these low carbon energy sources, their key advantages and limitations is shown in Table 6.5.

6.4 Next-Generation Energy Storage

With the rapid rise in the share of RE, energy storage is increasingly becoming important for the global energy transition. Power capacity is typically stated in GW, which measures the instantaneous rate at which a battery can supply electricity. GWh measures the total amount of energy stored in the battery, which can be delivered over a period of time. Both are important parameters as they decide the discharge duration of the storage system at full capacity.

Table 6.5 Advantages and limitations of low carbon energy sources

Energy source	Advantages	Limitations
Modern forms of bioenergy including solid biomass, biogas, biomethane, liquid biofuels	• Dispatchable and storable energy • Liquid biofuels suitable as drop-in fuels for transport • Net-negative emissions if accompanied by carbon capture and storage	• High water use • Competing use of land with food and ecosystems • Supply-chain emissions risk
Ocean energy including tidal stream, tidal range, wave, ocean thermal energy conversion (OTEC)	• High predictability • Minimal impact on land • Suitable for island communities	• Low technology readiness level (TRL) for some technologies • High costs and engineering challenges • Environmental permitting complexity
Waste-to-energy including municipal solid waste, landfill gas, sewage gas	• Lowers high GWP methane emissions • Small-scale and local energy production • Supports circular economy	• Limited feedstock availability • Emissions leakage concerns • Maintenance issues

$$\text{Discharge duration (hours)} = \text{Energy storage capacity (GWh)/Power capacity (GW)}$$

Energy storage systems having a response time of milliseconds and backup duration of several hours are required for ensuring grid flexibility, stability and higher share of RE integration into the grid. Long duration energy storage is also becoming increasingly important to provide daily and seasonal load balancing. As a rule of thumb for infrastructure planning, one unit of power capacity is required for every three units of renewable energy capacity added. This implies that 1 GW battery capacity is required for every 3 GW of variable renewable energy. Simulations are undertaken to account for variations due to demand profiles, technology, location and discharge time during the design stage of the energy system.

As shown in Fig. 6.1, global power capacity was about 474 GW at the end of 2024, with a larger share contributed by electrochemical batteries. However, pumped hydro systems provided the predominant share with 9000 GWh of energy storage capacity. Electrochemical batteries are rapidly being deployed for hourly storage, and there is also a transition from short-duration energy storage, which has a daily cycle, to long-duration energy storage for seasonal balancing.

It is estimated that to stay on the 1.5°-aligned path, about 1500 GW of storage capacity is required by 2030 [37]. Going further, 8000 GW of capacity would be needed by 2040 [43]. Nearly 200 countries pledged to work together to deploy 11,000 GW of RE capacity by 2030 at the COP 28 summit in Dubai, UAE [10]. At COP 29 in Baku, Azerbaijan, countries agreed on the 'Global Energy Storage and Grids Pledge' to collectively deploy 1,500 GW of energy storage capacity in the power sector by 2030 [11].

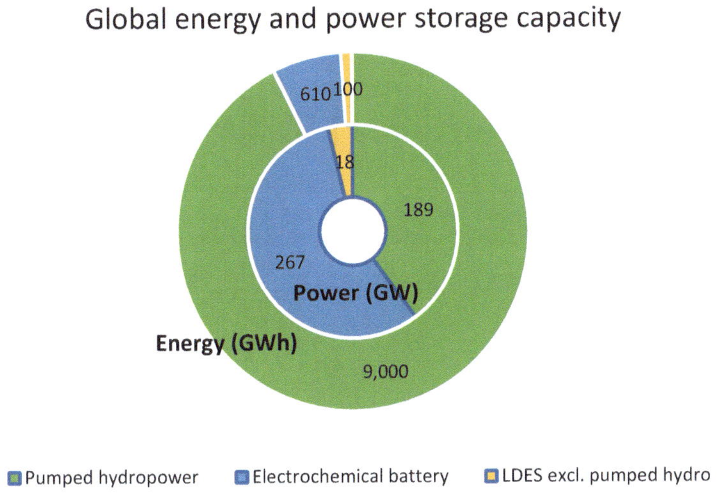

Fig. 6.1 Global energy storage and power capacity at the end of 2024. *Data source* IRENA Coalition for Action [41]

Energy can be stored in different forms such as chemical, electrical, mechanical, thermal, which can be converted to other forms for use. Table 6.6 shows a comparison of energy storage systems in terms of energy and power density, round-trip efficiency, response time, number of cycles, storage duration, operational life and the estimated cost. Table 6.7 shows the key advantages, limitations and primary applications of different types of energy storage technologies.

Steel and cement industries require high-temperature heat, which can be provided by molten salts that store renewable heat at temperatures greater than 1000° centigrade for industrial processes. Next-generation energy storage can also provide synthetic inertia along with response times in milliseconds for grid stability. These high-discharge and quick-response storage technologies can also provide peak load shaving, which avoids the need for peaking gas plants and the buildup of power supply infrastructure.

6.4.1 Advanced Batteries

Conventional batteries are undergoing an exponential cost decline. Costs have fallen by 99% over the past 30 years, with a 19% reduction for every doubling of deployment [65]. In 2025, costs fell to $108 per kilowatt-hour (kWh), which was less than half that in 2018 [46]. Energy density has increased fivefold over three decades, and the demand for Li-ion batteries reached nearly 1.6 TWh in 2025 [46]. Advanced battery technologies are moving away from critical minerals such as lithium and cobalt and

Table 6.6 Comparison of energy storage systems

Technology	Energy density	Power density	Efficiency	Response time	Cycle life	Storage duration	Operational life	Cost (2024–25)
Battery energy storage systems (BESS)—convert chemical energy directly into electrical energy through reversible reactions								
Lithium-ion batteries	150–300 Wh/kg	300–1500 W/kg	85–95%	< 1 s	3000–10000 cycles	1–4 h	10–15 yrs	$115–125/kWh
Flow batteries[a]	10–30 Wh/kg	~100 W/kg	75–90%	Seconds	>20 000 cycles	4–12 h	> 20 yrs	$300–600/kWh
Lead-acid batteries	30–50 Wh/kg	180–300 W/kg	70–85%	Seconds	500–1500 cycles	1–4 h	2–7 yrs	$50–100/kWh
Mechanical energy storage—convert electrical energy into mechanical potential or kinetic energy for storage								
Pumped hydroelectric storage	Low energy density	Low	70–85%	1–3 min	Decades	4–24 h+	50–100+ yrs	$1000–3000/kW
Compressed air energy storage[b] (CAES)	Low	Moderate	40–80%	Minutes	30–40 yrs	2–100 h	~30–50 yrs	$500–1500/kW
Flywheels	20–80 Wh/kg	1–5 kW/kg	85–95%	<1 s	>100,000 cycles	Seconds to minutes	Many years	$1500–3000/kW
Gravity energy storage[c]	0.05–1 Wh/kg	100–1,000 W/m^3	80–85%	Seconds	20,000–50,000 cycles	Hours to multi-day	30–50 years	$150–200/kWh
Thermal energy storage—stores heat/cold ideally for use in direct thermal applications in buildings and industrial processes								
Sensible heat (Molten salt)[d]	40–110 Wh/kg	10–100 W/kg	35–45% (thermal-to-electric)	Minutes-hours	10,000–30,000 cycles	6–15 h	10–30 yrs	$5–30/kWh (thermal)
Latent heat, phase change material (PCM)	50–150 Wh/kg[e]	10–100 kW/m^3	85–95% (heat use)	Minutes-hours	5,000–15,000 cycles	Hours to days	10–25 yrs	$10–50/kWh (thermal)

(continued)

Table 6.6 (continued)

Technology	Energy density	Power density	Efficiency	Response time	Cycle life	Storage duration	Operational life	Cost (2024–25)
Thermochemical storage[f]	100–400 kWh/m^3	10–200 kW/m^3	50–80% (system)	Minutes to hours	1,000–>10,000+ cycles	Hours to weeks/ seasonal	10–30 yrs	\$8–100/kWh (CapEx)
Electrical energy storage—stores energy directly in electrical or magnetic fields								
Supercapacitors/ ultracapacitors	1–10 Wh/kg	1–10 kW/kg	85–98%	Milliseconds	>1,000,000 cycles	Seconds to minutes	10+ yrs	\$5000–10,000/ kWh
Superconducting magnetic energy storage (SMES)[g]	0.5–5 Wh/kg	1,000–4,000 W/kg	> 95%	< 5 ms	Unlimited cycles	Seconds to minutes	30 yrs	\$1000–10,000/ kWh

Data source SolarTechOnline [71]

[a] E.g., Vanadium Redox flow batteries

[b] Diabatic CAES uses natural gas combustion during discharge; Adiabatic CAES stores compression heat for reuse; isothermal CAES maintains constant temperature during compression/expansion

[c] The system uses excess electricity to lift heavy masses and stores energy as gravitational potential energy. The mass is lowered when required to drive generators. Tower-based systems use 100 m+ tower with composite blocks; underground systems in mine shafts use suspended weights; mountain-based systems like 'Advanced Rail Energy Storage' uses rail cars on inclined tracks

[d] Example sodium and potassium nitrate

[e] Energy density for Paraffin waxes: 150–200 kJ/kg latent heat capacity; Salt hydrates: 100–300 kJ/kg capacity; Fatty acids: 180–250 kJ/kg. capacity; Metallic PCMs: 300–400 kJ/kg capacity

[f] Example calcium oxide/calcium hydroxide reactions and metal hydride systems

[g] SMES systems store energy in the magnetic field created by direct current flowing through a superconducting coil cooled below its critical temperature

Table 6.7 Advantages, limitations and applications of energy storage technologies

Technology	Advantages	Limitations	Primary applications
Battery energy storage systems (BESS)			
Lithium-ion batteries	High energy & power; fast response, modular	Safety/thermal risks; moderate life	Residential power backup, grid frequency regulation, peak shaving, RE firming
Flow batteries	Decoupling of power and energy; higher safety; deep discharge capability; easy recycling	Lower energy density; higher upfront cost	Utility-scale long-duration storage; microgrids; remote power supply
Lead-acid batteries	Low cost; mature tech; recyclable	Low life; high maintenance, heavy	Automotive applications; renewable energy storage; backup power supply
Mechanical energy storage			
Pumped hydroelectric storage	Very long life; High power capacity up to GW	Geographic constraints; high CapEx; environmental impact	RE integration; load balancing; ancillary grid services
CAES	Long duration; low marginal cost; Environmentally friendly black start capability	Lower efficiency; often geographic limits	Peak shaving; grid integration of RE; off-grid energy storage
Flywheels	High power density; long life span; minimal maintenance; high temperature range	High initial cost; high speed rotary equipment; self-discharge; space and weight constraints	Grid frequency regulation; peak shaving; backup power supply
Gravity energy storage	Low degradation; low operating and maintenance cost; environmentally friendly	Low energy density; high CapEx; limited capacity	Long-duration grid storage; Renewable firming; peak shaving
Thermal energy storage			
Sensible heat (Molten salt)	Low cost; long duration; low pressure	Large footprint; operational complexity, corrosive	CSP plants; industrial heat; power to heat
Latent heat (PCMs)	High energy density; isothermal operation; non-corrosive	Low thermal conductivity; high initial cost; high volume expansion	Thermal management in buildings; waste heat recovery in industrial processes; integration with renewable energy systems

(continued)

Table 6.7 (continued)

Technology	Advantages	Limitations	Primary applications
Thermochemical storage	Highly reversible reactions; high temperature potential; long-term low-loss storage	Low technical maturity, high cost, material stability issues	Seasonal storage in solar thermal plants; industrial heat; thermal integration with power cycles
Electrical energy storage			
Supercapacitors	Store energy directly in electrical or magnetic fields; operating temperature: −40 to +70 °C; minimal maintenance	High self-discharge	Electric vehicle regenerative braking systems; grid frequency regulation and voltage support; Power quality improvement; energy recovery in public transportation
SMES	Ultra-fast; high power capacity; no moving parts or chemical reactions	Very expensive; cryogenic cooling systems	Grid stability, magnetic levitation train systems, research facility power conditioning

are incorporating other battery chemistries to minimize the environmental impact. Some of the emerging battery technologies are briefly mentioned ahead.

6.4.1.1 Sodium-Ion Batteries

Sodium-ion batteries replace lithium with sodium, which is more abundant in the Earth's crust [47]. Sodium-ion batteries have high thermal stability and lower fire risks and could reduce material costs by about 25% as compared to lithium-ion batteries. Sodium extraction has a smaller ecological footprint than lithium refining and its wider availability reduces supply chain vulnerabilities. Although its energy density is lower than that of lithium-ion batteries, they can be deployed in small vehicles and stationary storage applications.

6.4.1.2 Solid-State Batteries

Solid-state batteries replace the flammable electrolyte with ceramics (oxides, sulfides, or phosphates), polymers, or composite materials. Their advantages include improvement in energy density over conventional lithium-ion batteries by 2–3 times, lower risk of fire, rapid charging capability, longer life due to a higher number of charging cycles and a wider operating temperature range. The technology is in early stages of commercial viability and the production cost at $400–800/kWh is several times higher than that of conventional batteries. Several automakers, including

Toyota, Honda and Mercedes-Benz, are planning a commercial launch between 2027 and 2030 [70].

6.4.1.3 Lithium-Sulfur Batteries

Lithium-sulphur batteries use sulfur as the cathode and lithium as the anode. Due to the battery chemistry, their energy density is about 10 times higher than traditional lithium-ion batteries. Sulfur is widely available and cheaper than nickel and cobalt, making the battery less expensive. Sulfur is nontoxic, has a lower environmental footprint and the battery is recyclable. These batteries are lighter and can be charged rapidly, but they have a shorter lifespan compared to lithium-ion batteries [32]. Companies like Stellantis and Zeta Energy are targeting deployment of these batteries in electric vehicles by 2028–2030 [19].

6.4.1.4 Flow Batteries

Flow batteries have a fundamentally different architecture, as energy is stored in liquid electrolyte solution, housed externally. Due to this design, the power and the energy storage capacity are independent of each other. The energy capacity of the battery depends on the size of the electrolyte tank, while the power of the battery is a function of the number and size of electrochemical cells. Flow batteries can scale up rapidly, have minimal performance loss and can last for longer lifetimes. As they use non-flammable liquid electrolytes, there is a smaller risk of fire, compared to lithium-ion batteries. Flow batteries are ideal for grid applications and renewable energy integration as they can store energy for 4–24 h. These are not dependent on critical minerals but use readily and commercially available materials. Common chemistries include vanadium-sulfuric acid, zinc-bromine and iron-chromium. Some of the largest flow batteries are deployed in utility-scale grid storage, as shown in Table 6.8.

An 800 MW/1600 MWh flow battery is under construction in Laufenburg, Switzerland. It is expected to commence operations in 2028 and will be the largest flow battery project in the world [69]. However, the 'Edwards & Sanborn' solar and energy storage project in California, which uses lithium-ion batteries with 3,287 MWh of energy storage capacity (operational Jan 2024) supplied by 120,000 battery units, would continue to remain the largest battery storage plant in the world [48].

6.4.2 *Battery Charging*

Along with developments in battery technologies, deployment of Electric Vehicle (EV) charging infrastructure and fast chargers has also grown rapidly. These are

Table 6.8 Examples of deployment of flow batteries

Name, location	Power, energy capacity	Operational	Role	Data source
Jimusar vanadium flow battery, Xinjiang (China)	200 MW power, 1000 MWh energy	Early 2026	Utility-scale storage integrated with 1 GW solar plant	[19]
Xinhua Ushi project, Xinjiang, (China)	175 MW power, 700 MWh energy	Dec 2024	Provides peak load support and frequency regulation for renewables	[18]
Dalian redox flow battery system, liaoning (China)	100 MW power, 400 MWh energy (plan to expand to 200 MW/800 MWh)	Oct 2022	Used for peak shifting and smoothing	[59]

essential for reducing charging time and help in the adoption of EVs, including buses, delivery vehicles and passenger cars. Large chargers, however, impact grid stability as it creates peak loads and may need to be paired with smart energy charging and energy storage devices. Standardization of chargers, bidirectional charging and improvements in battery charging software also lead to improved performance along with better operational efficiencies. A comparative assessment of different types of chargers on different parameters is shown in Table 6.9.

Some of the fastest charger-battery pairs include *CATL's* second-generation Shenxing superfast charging battery having an 800 km range. It has a peak charging power of 1.3 MW, and achieves 2.5 km of range per second of charging, adding 520 km of range in 5 min [8]. Similarly, in March 2025, *BYD Super e-Platform* achieved a peak charging speed of 2 km per second using peak charging power of 1 MW, with 5 min of charging, giving a range of 400 km [5].

Table 6.9 Comparative assessment of EV chargers

	Home charging	DC fast chargers	Next-generation extreme fast charging
Power output	3–20 kW (low voltage)	50–350 kW for small EVs; up to 4 MW for heavy-duty vehicles	350 kW to 1+ MW
Charging time	4–10 h for BEVs; 1–2 h for plug-in hybrid electric vehicle (PHEV)	20 min–1 h for 80% charge	10–15 min for 80% charge
Range added	40–75 miles in 1 h (for 11 and 22 kW charger, respectively)	60–200 miles in 20–30 min	400–520 miles in 5 min
Applications	Overnight, workplace charging and public charging facilities	Highway and commercial charging for long-distance travel	Future infrastructure for heavy-duty trucks and commercial vehicles

6.5 Emerging Transportation Technologies

6.5.1 Road Transport

Road transport decarbonization technologies that are being rapidly commercialized include Battery Electric Vehicles (BEVs), Plug-in Hybrid Electric Vehicles (PHEVs), and micromobility solutions such as e-bikes and e-scooters. Other emerging road transportation technologies are shown in Table 6.10.

6.5.2 Aviation

Use of Sustainable Aviation Fuel (SAF) derived from biomass, waste, or synthetic processes is growing in the aviation industry. Advanced Bio-SAF are produced from used cooking oil (UCO), animal fats (tallow), and vegetable oil through HEFA (Hydroprocessed Esters and Fatty Acids) processes. Other emerging aviation technologies having a high decarbonization potential are shown in Table 6.11.

6.5.3 Shipping

Liquefied Natural Gas (LNG) propulsion and dual fuel engines where ammonia and methanol are burned directly in modified engines are being used in ships. Over 500 ammonia and methanol-capable ships are on order and a few of them were operational in 2025. Battery-hybrid ferries and rigid wing and rotor sails technologies, which use modern wind propulsion for additional thrust or auxiliary power, are already being used onboard ships. Other emerging shipping technologies and fuels are shown in Table 6.12.

6.6 Industrial Decarbonization

In 2022, the industry sector emitted 9.0 Gt of CO_2 (including process emissions but not including indirect emissions from electricity for industrial processes), which was approximately one-fourth of global energy system CO_2 emissions [38].

Table 6.10 Emerging road transportation technologies

Technology	Principle	GHG reduction potential	TRL & time horizon	Key barriers	Use cases
Hydrogen fuel cell vehicles (FCVs)	Fuel cells convert hydrogen to electricity; have zero tailpipe emissions	60–95% (with green hydrogen)	TRL 7–8 (2028+)	Limited hydrogen infrastructure; high production costs	Heavy-duty long-haul transport where centralized refueling is available
Vehicle-to-grid (V2G) technology	Bi-directional electricity flow which allows EVs to supply power back to grid	Indirect impact; reduces quantity of electricity produced	TRL 7–8 (Pre-2030)	Battery degradation; requires smart grid infrastructure; regulatory issues	Areas with high EV penetration which also require grid stabilization
Electric road systems (ERS)	Charging and transfer of power via overhead lines, conductive rails, or in-road inductive coils	80–90% (with green power)	TRL 7 (2030+)	High upfront costs for infrastructure; lack of standardized interfaces	Urban transit, high-density freight corridors, autonomous shuttle fleets
Connected and autonomous vehicles (CAVs)	AI-driven vehicles with optimized routing and platooning	15–30% efficiency gains	TRL 6–7 (2030+)	Cybersecurity risks; high development costs; regulatory uncertainty	Dedicated freight corridors, ride-sharing fleets
Hydrogen internal combustion engine vehicle	Modified internal combustion engines using hydrogen	90%+ CO_2 reduction (with green hydrogen)	TRL 6 (2030–2040)	Limited refueling infrastructure, require high-pressure tanks, high NOx production	Heavy-duty long-haul trucking, construction, mining machinery

6.6.1 Low Emission Iron and Steel

In 2022, iron and steel contributed to 2.6 Gt of CO_2 emissions, which was 29% of the industrial emissions [36]. Table 6.13 lists the key technologies for steel decarbonization. Hydrogen Direct Reduced Iron (H2-DRI) is the most promising technology in the short term. The HYBRIT pilot plant in Luleå, Sweden (developed jointly by SSAB, LKAB and Vattenfall), which commenced operation in June 2021, was the

Table 6.11 Emerging aviation technologies

Technology	Principle	GHG reduction potential	TRL & time horizon	Key barriers	Use cases
Electric aircraft	Battery-electric propulsion using advanced lithium or solid-state batteries	90% (with green power)	TRL 7 (Pre-2030)	High battery weight; limited range; long recharge time	Short-haul (<400 km); island-hopping routes
Synthetic fuels such as e-SAF	E-fuels synthesized from captured CO_2 and green hydrogen	80–100%	TRL 6 (2030–40)	High cost (5–10 × jet fuel); energy intensive	Blending with SAF
Hydrogen fuel cell	Fuel cell converts hydrogen to electricity, which drives propellers	100% (with green hydrogen)	TRL 6 (2030+)	Requires cryogenic storage with 4 × tank space; high weight	Short-haul flights, small planes
Hybrid-electric propulsion	Combines conventional gas turbine engines with electric motors and batteries	30–50% fuel savings	TRL 5–6 (2035+)	Higher weight of dual systems; battery integration	Regional flights; cargo operations
Superconducting electric motors	High-temperature superconductors used in lightweight electric motors	80–100%	TRL 4–6 (2035+)	Cryogenic cooling required; high development cost	Regional routes with shorter range
Hydrogen-powered aircraft	Direct combustion of liquid hydrogen in jet engines	100% (with green hydrogen)	TRL 5 (2035+)	Requires cryogenic storage	Sub-2000 km cargo operations

world's first plant for production of hydrogen-reduced sponge iron. Other planned projects are shown in Table 6.14.

No large commercial Blast furnace (BF)—Carbon capture, utilization and storage (CCUS) plants are operational as yet. Since 2021, Tata Steel operates a carbon capture unit in its Jamshedpur plant in India which captures 5 tonnes of CO_2 per day. It uses an amine-based absorption process to capture carbon from blast furnace gas. The captured CO_2 is not stored, but is reused within the steel value chain. ArcelorMittal Gent, Belgium, along with partners Mitsubishi Heavy Industries, Ltd., is operating a pilot carbon capture unit with a capacity of 300 kg CO_2/day since May 2024.

Table 6.12 Emerging shipping technologies and fuels

Technology	Principle	GHG reduction potential (%)	TRL & time horizon	Key barriers	Use cases
Ammonia (NH_3) fuel cells	Solid oxide fuel cells (SOFC) using ammonia generate electricity for electric propulsion motors. Ammonia can also be 'cracked' into hydrogen for use in proton exchange membrane (PEM) fuel cells	80–95%	TRL 7 (2028+)	Ammonia requires to be stored in pressurized tanks or refrigerated tanks (at −33 °C); high toxicity and NO_x emissions; lower energy density	Deep-sea shipping, bulk carriers, container ships, future newbuilds
Methanol fuel cells	Direct methanol fuel cells (DMFC) to generate electricity for electric propulsion motors	70–90%	TRL 6 (Pre-2030)	Low energy density, toxic, corrosive, low flashpoint	Chemical tankers, harbor craft
Nuclear propulsion	Small modular reactors produce steam which power turbines	95%+	TRL 6–8 (Post-2035)	Port access risky; nuclear proliferation and safety concerns	Ice-breakers, deep-sea, Floating Power Units (FPUs)
Bio-LNG	Uses liquefied biomethane from waste in LNG powered ships	60–80%	TRL 8–9 (Pre-2030)	Methane leakage risk; feedstock is limited	Dual-fuel LNG vessels, ferries and short-sea shipping

Boston Metal commissioned the first multi-anode industrial MOE cell in 2025 at Woburn, Massachusetts, USA, and plans to build the first technology demonstration plant in 2026. BioIron™ is planning a pilot plant with Rio Tinto using raw biomass and microwave heating to produce low-carbon iron with a capacity of about 1 ton/hr in Western Australia, Australia, with a planned commissioning around 2026. Japan's COURSE50 (CO_2 Ultimate Reduction in Steelmaking by Innovative Technology for Cool Earth 50) program is developing top gas recycling in blast furnaces (TGR-BF) technology with CO_2 capture, and pilot projects are advancing, aiming at full industrial integration post-2030.

Table 6.13 Technologies for steel decarbonization

Technology	Description	GHG reduction potential	TRL and time horizon	Key barriers
Hydrogen-direct reduced iron (H2-DRI)—Electric arc furnace (EAF)	Uses green hydrogen as a reducing agent to convert iron ore to metallic iron, which is then melted in electric arc furnaces to produce steel	85–95% (with renewable electricity)	TRL 7–8 2028–2035	High cost of green hydrogen and low availability; RE and upgraded electrical grid; high-grade iron ore
Blast furnace with carbon capture and storage (BF-CCUS)	Post-combustion CO_2 capture from existing blast furnaces in integrated steel mills using scrubbing and separation technologies. Captured CO_2 is compressed for geological storage or utilization	40–60%	TRL 7–8 2028–2035	High CapEx costs; limited CO_2 storage capacity; retrofit challenging in existing plants
Molten oxide electrolysis (MOE)	Direct electrolytic reduction of iron ore dissolved in molten oxide electrolyte (1600 °C). Electric current splits iron oxide into iron and oxygen	95–100% (with renewable electricity)	TRL 4–6 2035–2045	Unproven at commercial scale; require ultra-high RE capacity; electrode durability challenges; high capital costs
Smelting reduction with biomass (Bio-SR)	Uses charcoal or biomass instead of metallurgical coal in iron reduction processes. Biomass provides carbon for reduction	60–80% (if biomass is sustainable)	TRL 7–8 2025–2030	Sustainable biomass supply; lower energy density; competing land use; applicable in a few countries like Brazil
Top gas recycling blast furnace (TGR-BF) with CCUS	BF injected with pure oxygen producing top gas with a very high concentration of CO_2 and reducing gases (CO and H_2). Captures and recycles top gas back into the furnace after removing CO_2 to act as a reductant for the iron ore. Captured CO_2 is used in chemical processes	25–50%	TRL 6–7 2028–2035	CCU market size is limited; Applicable to large blast furnaces

Table 6.14 Important H2-DRI steel decarbonization projects

Project name	Location	Capacity	Expected commencement
Baosteel Zhanjiang (China Baowu)	Guangdong, China	1.0 million tonnes per annum (TPA)	Production line connected in Dec 2025
Stegra	Boden, Sweden	2.5 million TPA; 5 million TPA (full capacity)	Phase 1 expected in 2026 Full-scale operations by 2030
tkH2Steel (thyssenkrupp Steel)	Duisburg, Germany	2.5 million TPA	2026–2027: commissioning with natural gas 2028: first injection of hydrogen planned 2029: Planned switch to 100% hydrogen operation
Vulcan Green Steel (Jindal Group)	Duqm, Oman	2.5 million TPA (two production lines planned, full capacity 5 million TPA)	2026–2027; initial plan to utilize natural gas
SALCOS® (Salzgitter AG)	Salzgitter, Germany	2.1 million TPA	2033–36

6.6.2 *Low-Carbon Cement and Concrete*

In 2022, cement contributed to 2.4 Gt of CO_2 emissions, which was 27% of the industrial emissions [36]. Table 6.15 lists the key technologies for cement decarbonization. Limestone Calcined Clay Cement (LC3) is the most promising technology in the short term for decarbonizing cement and concrete. In the long run, CCUS is a promising route for deep decarbonization as it reliably addresses process-related emissions. A combination of Supplementary Cementitious Materials (SCMs), which can deliver about 50% clinker reduction, followed by CCUS to produce near-zero emissions cement is a more practical approach. However, this approach depends on clay availability and access to CO_2 storage infrastructure.

Table 6.16 lists the important projects related to cement decarbonization. Other experimental-stage technologies include a project by *Sublime Systems* to develop a low-carbon, electrochemical process for producing cement at ambient temperatures which uses electricity to extract calcium from raw materials. A 30,000 Tons Per Annum (TPA) pilot facility in Holyoke, USA is planned for 2026 [77]. Geopolymers and alkali-activated materials include E-Crete™ made from fly ash (a byproduct of coal-fired power stations) and slag (a byproduct of steel manufacturing), which act as binders [86]. CarbonCure Technologies (Halifax, Canada) has produced over 38 million cubic meters of concrete using CO_2 mineralization and carbonation curing,

Table 6.15 Technologies for cement decarbonization

Technology	Description	GHG reduction potential	TRL and time horizon	Key barriers
Supplementary cementitious materials (Scms)—Calcined clay	Use of LC3 (Limestone Calcined Clay Cement) can replace Portland cement clinker. Clay is calcined at 750–850 °C (vs. 1450 °C for clinker), leading to lower process emissions and energy use	30–50% (with renewable heat)	TRL 8–9 2025–2030 for scaling	Standards harmonization; clay quality variability; clay deposits location
Geopolymers and alkali-activated materials	Use of alternative binders such as aluminosilicate materials (fly ash, slag, calcined clay, metakaolin) activated with alkaline solutions to form a binding gel. Limestone calcination is not required as polymerization occurs at lower temperatures (20–90 °C)	40–80% depending on feedstock and activator	TRL 7–8 2025–2030	Limited standards; fly ash availability declining; inconsistent feedstock quality
Carbon capture on cement kilns and storage	Captures CO_2 from cement kiln exhaust from both limestone calcination and fuel combustion using amine scrubbing followed by CO_2 storage or utilization	60–90%	TRL 7–8 2028–2035	CO_2 storage infrastructure; high capital cost; regulatory barriers; poor market adoption
CO_2 mineralization and carbonation curing	Injects CO_2 into fresh concrete during curing, where it forms calcium carbonate ($CaCO_3$), permanently sequestering CO_2	5–30% (optimized precast with mineralization)	TRL 7–8 2025–2030	Limited CO_2 supply infrastructure; poor CO_2 availability; lack of standards
Alternative binders	Use of Belite-Ye'elimite-Ferrite (BYF) and Climate-Smart Agriculture (CSA) during production lowers temperature for clinker production and uses less limestone	30–40% (with optimized formulations)	TRL 7–8 2025–2035	Bauxite availability and cost; standards still emerging
Electrified kilns	Uses electrical resistance heating, plasma technology, or induction heating	35–40% (with renewable electricity)	TRL 4–6 2030–2040	Requires process modifications; high electricity demand

Table 6.16 Important cement decarbonization projects

Technology	Project name (Company)	Location	Capacity	Status
SCMs/calcined clay	LC3 flash calciner (Heidelberg Materials and CBI Ghana)	Tema, Ghana	>400,000 t/yr calcined clay	Commenced production May 2025 [30]
Carbon capture on cement kilns and storage	Go4ECOPlanet Project (Holcim)	Kujawy, Poland	1.2 M t CO_2/yr capture; 2 M t cement	Targeted operations by 2030 [33]
Carbon capture on cement kilns and storage	Carbon2Business (Holcim)	Lägerdorf, Germany	1.2 M t CO_2/yr capture	Commenced Apr 2024; expected 2029
Carbon capture on cement kilns and storage	GO4ZERO Cement Plant (Holcim)	Obourg, Belgium	1.1 M t CO_2/yr capture	Commenced May 2024; expected 2029
Uses renewable electricity with carbon capture on cement kilns and storage	KOdeCO net zero project (Holcim)	Koromačno, Croatia	0.4 M t CO_2/yr capture	Expected 2028
Carbon capture on cement kilns and storage	CarboClearTech project (Holcim)	Martres-Tolosane, France	0.7 M t CO_2/yr capture	Expected 2031
CCUS with mineralization pathways	OLYMPUS project (Holcim)	Milaki, Greece	1.0 M t CO_2/yr capture; 2 M t cement	Commenced May 2025; expected operation 2029
Carbon capture on cement kilns and storage	Carbon Hub CPT01 (Holcim)	Câmpulung, Romania	2 M t cement	Expected operation 2032
Carbon capture on cement kilns and storage	Brevik CCS cement facility (Heidelberg)	Norway	400,000 t CO_2/yr	Operational June 2025 [31]

saving more than 500,000 metric tons of CO_2 to date across 7.5 million truckloads of CarbonCure concrete [7].

6.6.3 Low-Emission Chemicals and Plastics

Chemicals and plastics (including other petrochemicals) contributed to 1.3 Gt of CO_2 emissions, which was 15% of the industrial emissions in 2022 [36]. Table 6.17 lists some of the key technologies for low-emission chemicals and plastics.

Table 6.17 Technologies for low-emission chemicals and plastics

Technology	Description	GHG reduction potential	TRL and time horizon	Key barriers
Green ammonia production via electrolysis	Natural gas is replaced by hydrogen produced from RE in a steam methane reforming (SMR) based process. Hydrogen is then combined with nitrogen in the Haber-Bosch process, which is also powered by RE-based electricity	85–100% (with renewable electricity)	TRL 7–8 2025–2035	High cost; RE-based power availability
Bio-based polymers and chemical recycling for plastic production	Replaces fossil fuel as feedstock. Pathways include: (a) Direct fermentation of sugars to monomers/polymers (b) Fischer-Tropsch or bioethanol dehydration (c) Advanced recycling via pyrolysis or gasification to convert waste plastics back to monomers (d) Depolymerization (chemical or enzymatic) to virgin-quality monomers	Bio-based: 30–70% reduction Chemical recycling: 30–50% vs virgin plastic Mechanical recycling: 60–80% reduction	TRL 6–9 depending on technology 2025–2035	Sustainable feedstock availability; poor plastic collection infrastructure; waste contamination issues for recycling
Electrified steam crackers (E-cracking) for olefins production	Electric heating to crack hydrocarbons (naphtha, ethane, propane) through thermal decomposition. Methods involve resistance heating, plasma cracking, microwave-assisted cracking, or use of molten metal catalytic systems to produce ethylene, propylene, and byproducts	50–60% (with renewable electricity);	TRL 4–6 2030–2040	Materials durability; product quality; grid capacity; feedstock emissions unaddressed

(continued)

Table 6.17 (continued)

Technology	Description	GHG reduction potential	TRL and time horizon	Key barriers
E-Methanol to chemicals for production of olefins, aromatics, and chemical intermediates	Replaces fossil feedstocks with renewable carbon by producing green methanol from green hydrogen and captured CO_2 or biomass gasification	60–90%	TRL 6–9 2025–2035	High cost; limited CO_2 sources; other competing uses
Catalytic conversion of CO_2 to chemicals	Converts CO_2 to value-added chemicals through: (a) Electrochemical reduction using catalysts at moderate temperatures (b) Thermochemical conversion via reverse water–gas shift or fischer–tropsch process (c) Photocatalytic or photoelectrochemical conversion	30–90% depending on product, efficiency, and energy source	TRL 4–7 2028–2040	Low efficiency; high catalyst cost and low availability; limited high-purity CO_2 sources;

Some of the existing plants include green ammonia production since 2014 at Yara International's Heroya plant in Porsgrunn, Norway. The installed capacity is around 10,000 kg/d of hydrogen, which could produce 20,000 mt/year of ammonia, which can be converted to 60,000–80,000 mt/year of green fertilizer. This is planned to be expanded to 500,000 mt/year by 2030 [67]. NatureWorks plant in Blair, Nebraska, USA has a nameplate capacity of 150,000 metric tons of 'Ingeo' biopolymer since 2002 [53]. The George Olah plant by Carbon Recycling International in Svartsengi, Iceland was commissioned in 2012 and produces 4,000 metric tons of renewable methanol per year, capturing 5,500 tonnes of CO_2 annually from industrial emissions. Table 6.18 shows some low-emission chemicals and plastics projects under development.

6.6.4 *Low-Emission Textiles*

The textile industry contributed approximately 0.5% of CO_2 emissions in the industrial sector in 2022. The emissions were mainly from the use of fossil fuels for process heat for textile washing, steam generation and chemical-intensive wet processing (dyeing, finishing, scouring), and energy-intensive drying operations.

Table 6.19 lists important technologies for low-emission textiles.

Table 6.18 Important low-emission chemicals and plastics projects

Technology	Project Name	Location	Capacity	Status
Green ammonia production via electrolysis	NEOM Green Hydrogen Company	NEOM, Oxagon, Saudi Arabia	1.2 million tonnes/year of green ammonia; 600 tonnes per day of green hydrogen	80% construction completed as of mid-2025. Expected product availability in 2027 [54]
	Envision	Chifeng, China	320,000 tonnes/year green ammonia; planned expansion to 1.5 million tonnes/year	Operational in Q4 2025; Full capacity by 2028 [15]
Bio-based polymers and chemical recycling for plastic production	Emirates Biotech Polylactic Acid (PLA)[a] facility	United Arab Emirates	160,000 t / yr (two phases of 80,000 t each)	Expected operations by 2028 [13]
	Futerro PLA biorefinery	Normandy, France	200,000 t / yr combined lactic acid and PLA	Expected operations by 2029 [23]
E-cracking for olefins production	BASF-SABIC-Linde electrified steam cracking plant	Ludwigshafen, Germany	4 tons/h feedstock (demonstration plant)	Commenced production in 2024
	Technip Energies, LyondellBasell and Chevron Phillips Chemicals	Channelview, Texas, USA	Demonstration size	MoU signed in 2023 [78]
E-methanol to chemicals for production of olefins, aromatics, and chemical intermediates	Guangxi Huayi Energy Chemical MTO Plant	Guangxi, China	1,000,000 t/yr olefins	Completed in December 2025 (convertible to e-methanol feedstock use in future)

[a] PLA is biopolymer, is made from plants, is recyclable, compostable and biodegradable and can replace traditional plastics

Table 6.19 Technologies for low-emission textile production

Technology	Description	GHG reduction potential	TRL & time Horizon	Key barriers
Enzyme-based wet processing	Bio-catalytic enzymes (e.g., cellulases, amylases, laccases, pectinases, and proteases) replace or supplement harsh alkaline chemicals (NaOH, hydrogen peroxide, sulfuric acid) and reduce processing temperatures to 40–60 °C (conventional 90–95 °C+) in scouring, desizing, bleaching, and fabric finishing	20–40% thermal energy reduction with subsequent lowering of emissions Reduces chemical production and avoids emissions from wastewater treatment	TRL 8–9 Commercially deployable 2025 –2028	Enzyme stability is challenging at very high pH or temperature for some processes; cold storage requirements for specialty enzyme products High cost of enzymes
High-temperature industrial heat pumps	Heat pumps using high-temperature working fluids such as R744, R718 and ammonia use low-emission electricity to upgrade waste heat (40–80) to 160 °C	40–70% GHG reduction compared to gas boilers (100% decarbonization potential with zero-emission electricity)	TRL 7–9 Commercially scalable 2025 –2030	High capital cost; temperature output limited to 160–200 °C
Supercritical CO_2 dyeing	Supercritical CO_2 operating at 31 °C and 74 bar pressure replaces water as the dyeing solvent. The closed-loop system recycles CO_2 and residual dye	50–60% energy reduction compared to conventional water-based dyeing Emissions from wastewater treatment avoided	TRL 7–8 Commercial 2026–28, scale up beyond 2030	High capital cost for pressure vessels and CO_2 handling systems; significant process redesign; limited validated dye chemistry and materials

(continued)

Table 6.19 (continued)

Technology	Description	GHG reduction potential	TRL & time Horizon	Key barriers
Electrochemical indigo dyeing for denim and cotton warp yarn	Replaces sodium hydrosulfite (Na2S2O4) chemical reduction of indigo dye with electrochemical reduction at ambient temperature. Eliminates toxic reducing agents and associated waste streams	20–40% energy and Scope 1 emission reduction	TRL 7–8 Commercial deployment 2025–2030	High capital costs; complex retrofit to existing dyeing production lines; technology access limited
Low-temperature plasma treatment	Non-thermal plasma (NTP) activated by electrical discharge modifies fiber surfaces at ambient temperature, replacing energy- and chemical-intensive wet pre-treatment steps	Up to 30% pre-treatment energy and chemical production emissions Leads to lower dyeing bath temperatures, further reducing emissions	TRL 6–8 Commercial deployment 2027–2032	Current systems limited to batch-mode production; wash-fastness concerns; lack of standardized process protocols and industry testing standards
Green hydrogen for process heat	Electrolytic hydrogen produced from renewable electricity replaces fossil fuel boilers for high-temperature heat	100% reduction in Scope 1 emissions possible	TRL 5–7 2028–2035 for commercial scaling	High cost of green hydrogen; lack of hydrogen infrastructure in textile-producing countries; Safety regulations evolving

Enzyme-based denim biopolishing and desizing offered by Novonesis is used in several mills in China, Bangladesh, Pakistan, Turkey and India [55]. High-temperature industrial heat pumps manufactured by Mayekawa are used in demonstration projects to replace a propane-fired steam boiler operating in the food processing facility of Stella Polaris, Norway [29]. The heat pump uses water at 18 °C to produce steam at 150 °C and has applications in the textile industry. Supercritical CO_2 dyeing has been functional since 2019 with a Dyecoo Dyeox system in Ho Chi Minh City, Vietnam, with an operating capacity of ten tons per day and new projects are planned in several countries.

6.7 Agricultural and Biological Innovations

The agriculture sector contributed 6.2 $GtCO_2e$ GHG emissions, which was approximately 12% of global GHG emissions excluding Land Use, Land-Use Change, and Forestry (LULUCF) [9]. Methane (CH_4) emissions from enteric fermentation and rice cultivation and nitrous oxide (N_2O) emissions from fertilizer use are the main sources of emissions from the sector, other than land-use change.

6.7.1 *Methane-Reducing Feed Additives*

Enteric fermentation from global livestock is a large source of methane emissions. Additives like 3-Nitrooxypropanol (3-NOP) can be added to cattle feed, which inhibits methanogen enzyme activity in the rumen. Similarly, Bromoform, which is a natural, volatile organic compound produced by red seaweed halogenates blocks the enzymatic pathway, which reduces methane emissions from cattle. The GHG reduction potential varies between 20 and 80% enteric CH_4 reduction per animal depending on the product and application. Some of these products, like 3-NOP, are already commercially available in the EU, UK, Brazil, Australia and New Zealand. Trials of *Asparagopsis taxiformis* (seaweed) bromoform are underway, and it could be deployed commercially by the end of this decade. With methane emissions from livestock set to be taxed from 2030 in Denmark, use of these additives is likely to be adopted in certain countries. High cost, palatability, regulatory approval and availability of these additives in developing countries are some of the key barriers for scaling up deployment.

6.7.2 *Selective Breeding and Rumen Microbiome Engineering*

Biological approaches include genetic selection for breeding animals which naturally produce less enteric methane per unit of feed intake. It is estimated that 10–20% sustained methane reduction per animal can be obtained within 10 years of genetic selection with permanent progress. Selection of lower-emitting animals is being integrated into national livestock breeding programs in New Zealand, Ireland, Australia [80] and the UK, with larger deployment expected by the end of this decade.

Rumen microbiome engineering methods use targeted delivery of archaeal-suppressing viruses to infect and destroy methanogenic archaea (which are the methane producers in the rumen of cattle), with some studies claiming reductions of up to 97%. Anti-methanogen vaccines and next-generation direct-fed microbials are being designed to stimulate the immune system of the cattle to produce antibodies which enter the rumen and neutralize the targeted methanogens. Emission reduction of 10–30% in enteric methane is considered plausible [2]. Microbiome vaccines are still at trial stages and are likely to be deployed around the next decade. Challenges include long breeding cycles, complexity and time delays in regulatory pathways for approval of vaccines.

6.7.3 Enhanced Photosynthesis Engineering for Plants

Enhanced photosynthesis and metabolic engineering increases plant carbon fixation efficiency via genetic modification to lower the demand for nitrogen fertilizer. RuBisCO (Ribulose-1,5-bisphosphate carboxylase/oxygenase) is an enzyme which catalyzes the chemical reaction in plants to convert atmospheric CO_2. The enzyme is slow and inefficient and often reacts with oxygen instead of carbon dioxide through 'photorespiration', which wastes energy. Improving the catalytic rate of the RuBisCO enzyme, or introducing CO_2 concentrating mechanisms, can significantly reduce the oxygenation reaction and boost carbon fixation.

Crop engineering methods use multigene transformation to introduce specific photosynthesis traits found in crops like maize into crops like rice to minimize water loss and photorespiration. Photorespiratory bypass methods use alternative metabolic pathways to reduce the metabolic cost of photorespiration, which can boost yield by 20–40%. These methods can indirectly lead to lower land-use emissions and reduced N_2O emissions from the use of fertilizer. These technologies are at early stages of development and are likely to mature in the next decade. Regulatory hurdles for approvals, complex process of multigene editing and long R&D cycles are some of the challenges for these technologies.

6.7.4 Precision Fermentation

Precision fermentation uses genetically programmed microorganisms (yeasts, fungi, bacteria) in microbial bioreactors to produce functional ingredients that are structurally identical to livestock-based protein. Products include dairy proteins such as whey, egg white proteins and heme for plant-based meat. This can lead to avoided emissions from enteric methane production from animals, feed crop land use and manure management.

Precision fermentation dairy proteins can lower GHG emissions by 45–95% as compared to conventional dairy per kg protein. Products like dairy proteins and heme are closer to commercial deployment, while fats and complex lipids are likely to achieve market penetration by mid 2030s. Remilk and Gad Dairies in Israel are operating a fermentation facility to produce milk without cows since 2023 and plan to build the world's largest full-scale precision fermentation facility in Kalundborg, Denmark [62]. High production cost, long food regulation and approval processes and lower consumer acceptance are some of the challenges for the technology.

6.7.5 Carbon-Fixing Microorganisms

Soil microorganisms can be engineered using biological nitrogen fixation to replace synthetic fertilizers. This lowers the use of fertilizers, thereby reducing N_2O emissions. Examples include application of PROVEN®, a microbial soil amendment which has engineered strains of the bacterium Kosakonia Sacchari that colonize the root surfaces of corn plants. This fixes nitrogen directly from the atmosphere for supplying to the plant, reducing synthetic fertilizer demand [57].

Microbially mediated soil carbon sequestration technology includes biochar-microbiome inoculants where biochar is applied along with tailored microbial consortia to enhance soil carbon persistence and reduce N_2O emissions by 20–50% through improved nitrification/denitrification regulation. Methane-oxidizing bacteria (MOB) can also be used to oxidize methane before atmospheric emission. These measures can lead to N_2O reduction of 15–50% per hectare by replacing synthetic fertilizers. PROVEN® by 'Pivot Bio' is commercially available in the US; biochar-microbiome and MOB applied systems are expected to be deployed in the next 5–7 years. Challenges include field efficacy variability, impact of soil-microbiome interactions and poor adoption.

6.8 Technology for Nature-Based Solutions

Technology can enable Nature-based Solutions (NbS) which utilize forests, wetlands, and blue carbon to address climate change.

6.8.1 Advanced Land Monitoring Systems

Advanced monitoring systems use a combination of technologies such as ground-based IoT sensors to measure soil moisture and CO_2 flux, multispectral UAV surveys, and AI analytics to provide real-time, continuous data on carbon stocks and GHG fluxes.

Integrated platforms provided by 'Pachama' (bought by Carbon Direct) use satellite imaging with AI for remote carbon measurement and monitoring of forests. It also has automated carbon verification for reforestation projects in Latin America and Southeast Asia [6]. 'Global Forest Watch', an online platform by The World Resources Institute, provides forest data. It deploys near-real-time deforestation alerts globally using satellite data to monitor forests and helps to inform national REDD+ reporting [26].

Remote sensing from satellite constellations such as the European Space Agency's Sentinel missions provides Synthetic Aperture Radar (SAR) and multispectral data for earth observation [20]. Planet Labs' 'Dove', which provides 3–5 m resolution daily imagery, is being used for estimating scalable aboveground carbon and for rapid, high-frequency detection of forest disturbances [58]. NASA/USGS Landsat program provides long-term, continuous data for analyzing land use and to evaluate ecosystem, biodiversity and deforestation trends [49]. MAXAR's *WorldView* and *GeoEye* satellites provide high-resolution satellite imagery (30–1.2 m) and spectral analysis across visible, near-infrared and red-edge bands. This provides multispectral remote sensing data for environmental monitoring, localized analysis of forest cover change, and vegetation health monitoring [50].

Light Detection and Ranging (LiDAR) uses pulsed laser light to measure individual tree height, crown dimensions, and canopy density and can create a 3D model of forest canopy structure. This data can be used to convert aboveground biomass and carbon stock estimates at 10–30 m spatial resolution globally. Ground-based terrestrial laser scanning (TLS), drone-mounted sensors, airborne laser scanning (ALS) and NASA's spaceborne Global Ecosystem Dynamics Investigation (GEDI) LiDAR installed on the International Space Station have been used to develop forest maps [51].

The *Global Airborne Observatory* created 6.7 million hectare, 1.1 m resolution carbon stock maps for Peru using airborne LiDAR, which measured top-of-canopy height from the field and scaled the observations using machine learning. In Araguaia State Park, Brazil, high-density UAV-LiDAR was used to create detailed 3D land vegetation coverage models to estimate the amount of carbon stocks across forests, savannas, and grasslands [82].

6.8.2 Automated Reforestation Using Drone Seeding

Automated platforms that use GPS-guided mechanized planting equipment, drones, and AI-driven analysis are being used to restore degraded land. The process involves mapping and analysis using high-resolution satellite imagery to identify priority zones for seeding and to determine suitable native species. AI and machine learning algorithms are used to analyze several ecological variables such as soil composition, topography, microclimate, hydrology, and species physiology to create optimal planting patterns using automated tools. Purpose-built drones are used to disperse biodegradable seed pods containing germinated seeds, nutrients, and water-retention compounds inside protective covers. Post-deployment, satellite data monitors germination and growth rates.

Drones can deliver seed pods along a predefined route or can be deployed in a 'swarm', which can cover over 50 ha/day. This allows for cost-effective reforestation, especially in areas that are difficult to access. A single human supervisor can control and oversee the process, which makes it multiple times faster and cheaper than manual planting.

Flash Forest, a Canadian startup, uses drones equipped with pneumatic launchers and has planted several million trees across over 50 projects in wildfire-affected areas in Canada, the US, Australia and Brazil [21]. *ReForest Latam*, an Argentinian company, is also using AI screening with drone seeding and microorganism-treated iSeeds to accelerate reforestation across South America [61]. Another company, *Dendra Systems,* uses AI-driven drones and AI to restore ecosystems. Using high-resolution imagery, it precisely identifies plant species, assesses land health and uses drones to disperse seedpods, planting up to 700 kg of seeds per day. Automated reforestation has been used in over 20 ecosystem types, including projects for mangrove restoration in Abu Dhabi and koala habitat corridors in Australia [27].

Challenges include computational requirements for high-resolution modeling, high platform operation costs, limitations for high-quality field validation, seed dormancy issues, seed predation by birds leading to low germination rates in field conditions, short drone flight times and weight restrictions, and sourcing constraints for high-quality seeds.

6.8.3 Blue Carbon Technologies

Blue carbon refers to the carbon captured and stored by the world's coastal and marine ecosystems such as mangrove forests, salt marshes, and seagrass meadows. Technology can be used to conserve and restore these ecosystems, which boosts carbon sequestration in biomass and sediment.

Satellite imagery for coastal ecosystems allows scalable and precise mapping for mangroves. Hyperspectral imagery can be used to capture frequency bands beyond what is visible to the human eye and is effective at identifying vegetation species and health. LiDAR can be used to map the seabed to create accurate 3D maps of the ocean floor, including plant height and biomass, which can be used to estimate carbon stocks. Sonars can be used to provide valuable information on soil qualities including carbon value, as well as seabed topography and vegetation cover. This has potential to scale the estimation of seabed soil carbon stocks. Environmental DNA (eDNA) analyses the DNA released from organisms into the environment, and can be used for evidence-based biodiversity monitoring. eDNA data can be used to collect information on species presence, biodiversity estimates, as well as information about ecosystem dynamics [3].

Mangroves, saltmarshes, and seagrass meadows can sequester five to 30 times more carbon underground per unit area than terrestrial ecosystems and their conservation and restoration can provide higher value per dollar of investment [45]. The Environment Agency—Abu Dhabi (EAD) planted one million mangrove seeds using drones in the first phase under the Abu Dhabi Mangrove Initiative [14]. An agreement to plant an additional 2.5 million mangrove seedlings across Abu Dhabi using drones has been signed with a further commitment to plant up to 10 million mangroves in Abu Dhabi by 2030 [1].

Seagrass can trap organic material and sediment in its roots, making it effective for climate mitigation. Automated seagrass restoration is undertaken using robots, unmanned underwater vehicles (UUVs), and mechanized systems. Seeds are harvested from donor meadows and underwater robots are used to plant them in a standardized, automated manner. 'Ulysses Ecosystem Engineering' uses UUVs navigated by sonar and 3D vision to harvest, plant, and monitor seagrass with high precision. It can plant five hectares of seagrass per day [79]. The Automated Seagrass Planter uses ruggedized electronics to ensure optimal spacing of biodegradable seed-pods and seagrass plugs and records planting locations via the Global Positioning System (GPS).

Commercial scale-up of technologies faces significant technical monitoring challenges and cost hurdles. Environmental challenges include slow pace of natural restoration, sensitivity to water parameters, permanence risks from sea-level rise and storms, undeveloped carbon accounting protocols and marine governance issues in coastal zones.

6.9 Solar Radiation Modification

Geoengineering methods use technology for interventions in the Earth's climate system to address climate change. Solar Radiation Modification (SRM) is a solar geoengineering method which uses a suite of technologies to reflect a portion of sunlight back into space. It is estimated that reflecting 1% more sunlight would offset around 1 °C of global warming, which leads to a large interest in these methods. However, SRM does not address increases in GHG concentration and must be seen as a transitional intervention which can complement GHG emission reduction. Some of the proposed technologies are mentioned ahead.

6.9.1 Stratospheric Aerosol Injection

Stratospheric aerosol injection (SAI) methods inject reflective aerosols such as sulphur dioxide (SO_2) or calcium carbonate ($CaCO_3$), 20–25 km above the Earth's surface using high-altitude aircraft, stratospheric balloons, and tethered hoses. These aerosols scatter incoming shortwave solar radiation back into space, thereby artificially suppressing surface temperature temporarily. While sustained use of SAI could offset 1–2 °C of global warming, the technology carries critical risks such as rapid rebound in global warming upon ending its usage and asymmetric impacts on precipitation.

The Stratospheric Controlled Perturbation Experiment (SCoPEx) research programme [42] at Harvard University aimed to measure the impact of SAI and to advance scientific knowledge relevant to potential future implementation of solar geoengineering. The first flight was postponed following stakeholder concerns, and further work has been suspended [68]. Another experiment under the Stratospheric Aerosol Transport and Nucleation (SATAN) project was conducted in the UK in 2022 as an engineering proof of concept. It released about 400 g of sulphur dioxide into the stratosphere to examine aerosol microphysics and transport dynamics relevant to SAI efficacy [72].

While laboratory and controlled outdoor experiments have been conducted, full-scale field deployment is decades away due to lack of governance and an international legal framework for application of these technologies, concerns about its impact on weather systems and possible reactions which could lead to ozone depletion in the stratosphere.

6.9.2 Surface Albedo Modification

Surface albedo modification is predicated on increasing the reflectivity of Earth's surface. Different surfaces reflected different proportions of sunlight, which range from 12% for worn asphalt, 6% for oceans, 8–18% for forests, 40% for deserts, and up to 90% for snow and ice. Concepts for increasing the reflectivity of Earth's surfaces have been suggested [73]. Painting roofs and pavements with white colour can reduce surface temperatures by several degrees as it reflects sunlight back into space, but there are logistical and ecological challenges for large-scale deployment. Proposed methods also include covering Arctic ice with reflective materials to slow melting. Surface albedo modification can help produce local coolingeffects, but its potential to lower global temperature is limited.

6.9.3 Marine Cloud Brightening

Marine Cloud Brightening (MCB) involves spraying fine sea-salt aerosols into low-level marine stratocumulus clouds. This increases the number of cloud condensation nuclei, leading to smaller, more numerous droplets. A larger number of smaller droplets makes a cloud appear brighter and whiter, thus reflecting more sunlight back into space. modeling suggests a regional cooling impact of 0.5–1 °C from intensive deployment of MCB in major stratocumulus regions. Similar to other SRM methods, the method does not reduce atmospheric CO_2 stock but provides a temporary cooling effect which is likely to be reversible on a short timescale.

MCB has been tested in the Great Barrier Reef, Australia since 2020, where sea-salt mist generators were deployed to deliver tiny sea-salt particles to the lower atmosphere from a ship. In 2024, an airborne platform was used to test how the aerosol particles affect low-level clouds. Results showed localised cloud brightening with limited efficacy, which suppressed rain [74]. The University of Washington (USA), Marine Cloud Brightening Project also conducted small-scale field studies involving a single plume of generated sea salt aerosol in the northeast Pacific [81].

As there have been only controlled field experiments, there is limited data on the impact and the technology will take decades to mature. Concerns include marine ecosystem impacts, unequal regional effects on precipitation, predictability issues due to nonlinear impacts, lack of a science-based governance system for MCB deployment and slow regulatory framework development.

6.9.4 Space-Based Solar Shields

Space-based solar shields are a theoretical concept that proposes the deployment of structures in space to block or deflect a portion of the Sun's radiation before it reaches Earth. These shields aim to reduce solar insolation and do not lower GHG emissions. Solar shields made from thin-film bubbles, swarms of autonomous reflective spacecraft or lunar dust injection at 1.5 million kilometers from Earth have been proposed. A 1.7% reduction in solar insolation is calculated to offset approximately 1.7 °C of global warming, showing high theoretical potential for the concept. However, the cost is potentially estimated in the range of 1–5 trillion dollars and at the moment there are no credible deployment pathways [75]. The environmental risks and the long-term ecological consequences of modifying solar insolation reaching the Earth are also not fully understood.

6.9.5 Cirrus Cloud Thinning

Cirrus Cloud Thinning (CCT) aims at reducing the warming effect of high-altitude (5–10 miles) cirrus clouds, which are known to trap outgoing infrared radiation (heat) from the Earth's surface. Seeding these clouds with ice-nucleating particles such as bismuth-containing substances creates larger ice crystals which fall faster, reducing the cloud's lifespan and optical depth. This reduces the top-of-atmosphere net cloud radiative effect, allowing the heat to escape into space, causing a cooling effect. There is low confidence in this method due to limited understanding of cirrus microphysics and aerosol interactions. Large uncertainties in the feasibility and high risks associated with this method place this idea at a concept stage with no practical ways proposed for its implementation [76].

6.10 Conclusion

Technology is a critical enabler of climate action. Achieving global net-zero emissions will require a portfolio of technologies deployed across several sectors such as power, energy storage, transport, industry and agriculture. Electrification is an important component of decarbonization and a combination of renewable electricity generation, end-use electrification, modern power grids, demand-side flexibility, digitalisation, and artificial intelligence will help advance low-carbon climate solutions.

Technological advances in solar PV, floating offshore wind, advanced nuclear reactors and novel battery chemistries are becoming drivers of emerging climate solutions as they become more efficient, cheaper and user-friendly. Technologies are rapidly evolving and advanced solutions are being deployed in road transportation, aviation and shipping sectors. Emerging technologies for low-emission steel, cement, chemicals, plastics, textiles, and decarbonisation of agriculture are becoming technically feasible, although large-scale deployment is currently challenging due to the high cost and low TRL. Technological innovation can accelerate the deployment of nature-based solutions through advanced monitoring systems, automated seeding using drones and blue carbon solutions. Emerging options such as solar radiation modification require additional scientific research, strong governance, and risk assessment before it is considered for deployment.

New technology needs to be commercialized and scaled, which is dependent on lower technology cost, supportive policies such as incentives, infrastructure, skilled workforce, manufacturing capacity, markets, public acceptance and production rights for developing countries. Successful deployment of climate solutions will depend not only on breakthrough technologies but also on rapidly scaling the existing proven solutions that translate into measurable climate impact.

References

1. ADNOC (2023) ADNOC uses innovative drone technology to plant mangroves. https://www.adnoc.ae/en/news-and-media/press-releases/2023/adnoc-uses-innovative-drone-technology-to-plant-mangroves. Accessed 18 Feb 2026
2. Ag Matters (no date) Future actions. https://www.agmatters.nz/actions/future-actions/. Accessed 30 Jan 2026
3. Blue Marine Foundation (2025) Innovative technology in blue natural capital projects. https://www.bluemarinefoundation.com/wp-content/uploads/2025/06/Innovative-Technology-in-Blue-Natural-Capital-Projects-v1.pdf. Accessed 18 Feb 2026
4. Buljan A (2024) PHOTO: Mingyang's 16.6 MW dual-turbine floating wind platform enters operation in China. Offshore Wind. https://www.offshorewind.biz/2024/12/12/photo-mingyangs-16-6-mw-dual-turbine-floating-wind-platform-enters-operation-in-china/. Accessed 15 Jan 2026
5. BYD (2025) BYD unveils super e-platform with megawatt flash charging for electric vehicles, matching refueling speeds. https://www.byd.com/en/news-list/BYD-Unveils-Super-e-Platform-Megawatt-Flash-Charging-Electric-Vehicles-Matching-Refueling-Speeds.html. Accessed 18 Jan 2026
6. Carbon Direct (no date) Homepage. https://www.carbon-direct.com/. Accessed 10 Feb 2026
7. CarbonCure Technologies (2024) CarbonCure achieves milestone half million metric tons of CO_2 savings. https://www.carboncure.com/news/carboncure-achieves-half-million-metric-tons-of-carbon-savings/. Accessed 23 Jan 2026
8. CATL (no date) Naxtra battery breakthrough & dual-power architecture: CATL pioneers the multi-power era. https://www.catl.com/en/news/6401.html. Accessed 18 Jan 2026
9. Climate Watch (World Resources Institute) (no date) Agriculture. https://www.climatewatchdata.org/sectors/agriculture. Accessed 30 Jan 2026
10. COP28 Presidency (2023) Global renewables and energy efficiency pledge. https://www.cop28.com/en/global-renewables-and-energy-efficiency-pledge. Accessed 14 Jan 2026

11. COP29 Presidency (2024) Global energy storage and grids pledge. https://cop29.az/en/pages/cop29-global-energy-storage-and-grids-pledge. Accessed 14 Jan 2026
12. EDF Group (2025) Provence grand large: full commissioning of the first french floating offshore wind farm. https://www.edf.fr/en/the-edf-group/dedicated-sections/journalists/all-press-releases/provence-grand-large-full-commissioning-of-the-first-french-floating-offshore-wind-farm. Accessed 15 Jan 2026
13. Emirates Biotech (2024) Emirates biotech selects Sulzer technology to build world's largest polylactic acid production facility. https://emiratesbiotech.com/emirates-biotech-selects-sulzer-technology-to-build-worlds-largest-polylactic-acid-production-facility/. Accessed 26 Jan 2026
14. Environment Agency—Abu Dhabi (2023) As part of Abu Dhabi Mangrove initiative, Environment Agency—Abu Dhabi Plants One Million Mangrove Seeds by Drone. https://www.ead.gov.ae/en/Media-Centre/News/0-plants-one-million-mangrove. Accessed 27 June 2026
15. Envision (2025) Envision delivers on world's largest green hydrogen and ammonia plant with off-grid renewable system. https://www.envision-group.com/news-detail?id=6881eccce98bb4003597fdf6&lang=en. Accessed 26 Jan 2026
16. Equinor (no date) Hywind Tampen. Available at: https://www.equinor.com/energy/hywind-tampen (Accessed: 15 January 2026).
17. Equinor (no date) Hywind Scotland. https://www.equinor.com/energy/hywind-scotland. Accessed 15 Jan 2026
18. ESS News (2024) World's largest vanadium redox flow project completed. https://www.ess-news.com/2024/12/09/worlds-largest-vanadium-redox-flow-project-completed. Accessed 17 Mar 2026
19. ESS News (2026) China connects world's largest vanadium flow battery project. https://www.ess-news.com/2026/01/07/china-connects-worlds-largest-vanadium-flow-battery-project/. Accessed 17 Mar 2026
20. European Space Agency (no date) The sentinel missions. https://www.esa.int/Applications/Observing_the_Earth/Copernicus/The_Sentinel_missions. Accessed 10 Feb 2026
21. FlashForest (no date) Homepage. https://www.flashforest.com/. Accessed 18 Feb 2026
22. Flotation Energy (no date) Kincardine. https://flotationenergy.com/kincardine-once-the-worlds-largest-floating-wind-project/. Accessed 15 Jan 2026
23. Futerro (no date) Futerro officially submits permit applications for its future biorefinery in Port-Jérôme-sur-Seine. https://www.futerro.com/futerro-submits-permit-applications-for-its-future-biorefinery/. Accessed 26 Jan 2026
24. Generation IV International Forum (no date) Welcome to the generation IV International Forum. https://www.gen-4.org/. Accessed 10 Jan 2026
25. Global Energy Monitor (no date) Hainan Wanning offshore floating wind farm. https://www.gem.wiki/Hainan_Wanning_Offshore_Floating_wind_farm. Accessed 05 Jan 2026
26. Global Forest Watch (World Resources Institute) (no date) Forest monitoring, land use & deforestation trends. https://www.globalforestwatch.org/. Accessed 10 Feb 2026
27. Global Society (no date) Satellite-guided drone reforestation: a high-tech solution to climate and biodiversity crises. https://www.globalsociety.earth/post/satellite-guided-drone-reforestation-a-high-tech-solution-to-climate-and-biodiversity-crises. Accessed 18 Feb 2026
28. GreenMatch (no date) Efficiency of perovskite solar cells. https://www.greenmatch.co.uk/blog/perovskite-solar. Accessed 04 Jan 2026
29. Heat Pumping Technologies (2025) Industrial high-temperature heat pumps: industrial heat pump demonstration case—Shrimp process plant. Project 68. https://heatpumpingtechnologies.org/content/uploads/sites/81/2026/01/demo14spiritmayekawa20251107.pdf. Accessed 27 June 2026
30. Heidelberg Materials (2025) Scaling up low-carbon clinker alternatives: world's largest calcined clay plant in Ghana starts production. https://www.heidelbergmaterials.com/en/pr-2025-05-16. Accessed 23 Jan 2026
31. Heidelberg Materials (2025) World premiere at Heidelberg materials: opening of CCS facility in Norway marks new era of sustainable construction. https://www.heidelbergmaterials.com/en/pr-2025-06-18. Accessed 23 Jan 2026

32. Highstar (2025) Lithium-sulfur battery: the next generation of energy storage technology. https://en.highstar.com/blog/lithium-sulfur-battery-next-generation-energy-storage. Accessed 15 Jan 2026
33. Holcim (no date) Go4ECOPlanet. https://www.holcim.com/what-we-do/green-operations/ccus/go4ecoplanet. Accessed 23 Jan 2026
34. Interesting Engineering (2025) The design behind the world's most efficient 27.81% silicon cell. https://interestingengineering.com/science/most-efficient-solar-cell-explained. Accessed 04 Jan 2026
35. International Atomic Energy Agency (2024) Small modular reactors: advances in SMR developments 2024. IAEA, Vienna. https://www-pub.iaea.org/MTCD/Publications/PDF/p15790-PUB9062_web.pdf. Accessed 05 Jan 2026
36. International Energy Agency (2022) Direct CO_2 Emissions from industry in the net zero scenario, 2000–2030. Chart. https://www.iea.org/data-and-statistics/charts/direct-co2-emissions-from-industry-in-the-net-zero-scenario-2000-2030-2. Accessed 18 Jan 2026
37. International Energy Agency (2024) Batteries and secure energy transitions. IEA, Paris. https://www.iea.org/reports/batteries-and-secure-energy-transitions/executive-summary. Accessed 14 Jan 2026
38. International Energy Agency (no date) Industry. Energy System. https://www.iea.org/energy-system/industry. Accessed 18 Jan 2026
39. IRENA (2024) Floating offshore wind outlook. International Renewable Energy Agency, Abu Dhabi. https://www.irena.org/-/media/Files/IRENA/Agency/Publication/2024/Jul/IRENA_G7_Floating_offshore_wind_outlook_2024.pdf. Accessed 05 Jan 2026
40. IRENA (2025) Renewable power generation costs in 2024. International Renewable Energy Agency, Abu Dhabi. https://www.irena.org/-/media/Files/IRENA/Agency/Publication/2025/Jul/IRENA_TEC_RPGC_in_2024_2025.pdf. Accessed 04 Jan 2026
41. IRENA Coalition for Action (2025) Key enablers for the energy transition: solar and storage. https://coalition.irena.org/-/media/Files/IRENA/Coalition-for-Action/Publication/IRENA_PAR_Coalition_Key-Enablers-Energy-Transition--Solar-and-Storage_2025.pdf. Accessed 14 Jan 2026
42. Keutsch Group (no date) SCoPEx. https://www.keutschgroup.com/scopex. Accessed 23 Feb 2026
43. LDES Council (2024) 2024 annual report. https://ldescouncil.com/wp-content/uploads/2025/05/2024LDESAnnualReport-1.pdf. Accessed 14 Jan 2026
44. LONGi (no date) 34.85%! LONGi breaks world record for crystalline silicon-perovskite tandem solar cell efficiency again. https://www.longi.com/en/news/silicon-perovskite-tandem-solar-cells-new-world-efficiency/. Accessed 04 Jan 2026
45. maribus (2024) World ocean review 8: the ocean—A climate champion? How to boost marine carbon dioxide uptake. maribus gGmbH, Hamburg. https://worldoceanreview.com/wp-content/downloads/wor8/WOR8_en_chapter_5.pdf. Accessed 25 Jan 2026
46. McKinsey (2026) Battery 2035: building new advantages. https://www.mckinsey.com/features/mckinsey-center-for-future-mobility/our-insights/battery-2035-building-new-advantages. Accessed 25 Mar 2026
47. MIT Technology Review (2026) Sodium-ion batteries: 10 breakthrough technologies 2026. https://www.technologyreview.com/2026/01/12/1129991/sodium-ion-batteries-2026-breakthrough-technology/. Accessed 20 June 2026
48. Mortenson (no date) Edwards & Sanborn solar + energy storage. https://www.mortenson.com/projects/edwards-sanborn-solar-plus-storage. Accessed 17 Mar 2026
49. NASA (no date) Landsat. NASA science. https://science.nasa.gov/mission/landsat/. Accessed 15 Feb 2026
50. NASA Earthdata (no date) LA Fires Maxar. Data Catalog. https://www.earthdata.nasa.gov/dashboard/data-catalog/la-fires-maxar. Accessed 15 Feb 2026
51. NASA Earthdata (no date) Global ecosystem dynamics investigation Lidar. https://www.earthdata.nasa.gov/data/instruments/gedi-lidar. Accessed 15 Feb 2026

52. National Rural Utilities Cooperative Finance Corporation (2025) A glimpse at the future of space-based solar power. https://www.nrucfc.coop/content/solutions/en/stories/energy-tech/a-glimpse-at-the-future-of-space-based-solar-power.html. Accessed 05 Jan 2026
53. NatureWorks (no date) About NatureWorks. https://www.natureworksllc.com/about-natureworks. Accessed 23 Jan 2026
54. NEOM Green Hydrogen Company (no date) Homepage. https://nghc.com/. Accessed 26 Jan 2026
55. Novonesis (no date) Enzymes and solutions for industrial bioprocessing. https://www.novonesis.com/en/biosolutions/industrial-bioprocessing/textile. Accessed 23 Jan 2026
56. Ocean Winds (no date) WindFloat Atlantic. https://windfloat-atlantic.com/. Accessed 15 Jan 2026
57. Pivot Bio (no date) PROVEN 40: a microbial fertilizer for corn. https://www.pivotbio.com/corn. Accessed 10 Feb 2026
58. Planet (no date) Forest carbon: reliable, scalable carbon estimates. https://www.planet.com/products/forest-carbon/. Accessed 15 Feb 2026
59. PV Magazine (2022) China connects world's largest redox flow battery system to grid. https://www.pv-magazine.com/2022/09/29/china-connects-worlds-largest-redox-flow-battery-system-to-grid/. Accessed 17 Mar 2026
60. PV Tech (2024) China's CHN energy completes world's largest open sea floating solar PV project. https://www.pv-tech.org/chinas-chn-energy-completes-worlds-largest-open-sea-floating-solar-pv-project/. Accessed 05 Jan 2026
61. Reforest (no date) Homepage. https://reforest-latam.com/. Accessed 18 Feb 2026
62. Remilk (2022) Remilk to build the world's largest precision fermentation facility. https://www.remilk.com/newsroom/remilk-to-build-the-worlds-largest-precision-fermentation-facility. Accessed 30 Jan 2026
63. Reuters (2025) Google announces Tennessee as site for small modular nuclear reactor. https://energynow.com/2025/08/google-announces-tennessee-as-site-for-small-modular-nuclear-reactor/. Accessed 10 Jan 2026
64. Reuters (2025) Sweden's Vattenfall shortlists Rolls-Royce, GE Vernova to build SMR nuclear reactors. https://www.arctictoday.com/swedens-vattenfall-shortlists-rolls-royce-ge-vernova-to-build-smr-nuclear-reactors/. Accessed 10 Jan 2026
65. RMI (no date) The rise of batteries in six charts and not too many numbers. https://rmi.org/the-rise-of-batteries-in-six-charts-and-not-too-many-numbers/. Accessed 15 Jan 2026
66. Rosatom Newsletter (2024) Atoms grow in appeal. https://rosatomnewsletter.com/2024/01/27/atoms-grow-in-appeal/. Accessed 10 Jan 2026
67. S&P Global (2024) Yara starts up Europe's largest green hydrogen plant in Norway. https://www.spglobal.com/energy/en/news-research/latest-news/energy-transition/061024-yara-starts-up-europes-largest-green-hydrogen-plant-in-norway. Accessed 25 Jan 2026
68. Salata Institute for Climate and Sustainability, Harvard University (no date) An Update on SCoPEx. https://salatainstitute.harvard.edu/an-update-on-scopex/. Accessed 23 Feb 2026
69. SCCIJ News (2025) Switzerland to operate the world's largest storage battery. https://sccij.jp/news/switzerland-to-operate-the-worlds-largest-storage-battery/. Accessed 17 Mar 2026
70. SolarTechOnline (2025) Solid state batteries: complete guide to technology, benefits & timeline. https://solartechonline.com/blog/solid-state-batteries-complete-guide/. Accessed 15 Jan 2026
71. SolarTechOnline (2026) Types of energy storage technologies: complete guide. https://solartechonline.com/blog/types-of-energy-storage-technologies/. Accessed 27 Mar 2026
72. SRM360 (no date) Outdoor experiments: SATAN. https://srm360.org/outdoor-experiments/satan/. Accessed 23 Feb 2026
73. SRM360 (no date) Surface Albedo modification. https://srm360.org/article/surface-albedo-modification/. Accessed 23 Feb 2026
74. SRM360 (no date) Outdoor experiments: reef restoration and adaptation project (RRAP) marine cloud brightening. https://srm360.org/outdoor-experiments/reef-restoration-and-adaptation-project-rrap-mcb/. Accessed 23 Feb 2026

75. SRM360 (no date) Why research space-based SRM? https://srm360.org/perspective/why-research-space-based-srm/. Accessed 25 Feb 2026
76. SRM360 (no date) Cirrus cloud thinning. https://srm360.org/article/cirrus-cloud-thinning/. Accessed 25 Feb 2026
77. Sublime Systems (no date) Manufacturing process. https://sublime-systems.com/our-process/. Accessed 23 Jan 2026
78. Technip Energies (2023) Technip energies, LyondellBasell and Chevron Phillips Chemical Sign MOU for Electric Cracking Ethylene Furnace. https://www.ten.com/en/media/news/technip-energies-lyondellbasell-and-chevron-phillips-chemical-sign-mou-electric-cracking. Accessed 26 Jan 2026
79. Ulysses (no date) Use cases: nature. https://www.ulysses.inc/use-cases/nature. Accessed 23 Feb 2026
80. University of New England/Animal Genetics and Breeding Unit (no date) BREEDPLAN. https://breedplan.une.edu.au/. Accessed 30 Jan 2026
81. University of Washington (no date) Marine cloud brightening program. https://atmos.uw.edu/faculty-and-research/marine-cloud-brightening-program/. Accessed 25 Feb 2026
82. Woodwell Climate Research Center (no date) LiDAR technology and carbon biomass estimates. https://www.woodwellclimate.org/lidar-technology-carbon-biomass-estimates/. Accessed 15 Feb 2026
83. World Nuclear Association (no date) Shidaowan HTR-PM 1. Reactor database. https://world-nuclear.org/nuclear-reactor-database/details/Shidaowan-HTR-PM-1. Accessed 10 Jan 2026
84. World Nuclear Association (no date) Generation IV nuclear reactors. https://world-nuclear.org/information-library/nuclear-power-reactors/other/generation-iv-nuclear-reactors. Accessed 10 Jan 2026
85. World Nuclear News (no date) China's demonstration HTR-PM enters commercial operation. https://www.world-nuclear-news.org/Articles/Chinese-HTR-PM-Demo-begins-commercial-operation. Accessed 10 Jan 2026
86. Zeobond (no date) E-Crete geopolymer concrete, fly ash concrete. http://www.zeobond.com/products-e-crete.html. Accessed 23 Jan 2026

Chapter 7
New Frontiers in Climate Action

Abstract This chapter focuses on emerging frontiers and identifies the drivers of the new phase of climate action. It spotlights Artificial Intelligence (AI) as an enabler for climate solutions and discusses its applications, including renewable energy integration, research for carbon capture materials, electric vehicles, batteries, smart buildings, urban planning solutions, climate monitoring and weather prediction. Open-source climate models, climate data platforms, climate risk mapping and climate decision support tools for accelerating climate action are presented as applications of digital public goods. It discusses how individual and corporate venture capital, private equity, angel investors, sovereign wealth funds and family offices can contribute to advancing climate tech. The chapter concludes by emphasizing the need for developing a system approach for effective climate action.

Keywords Artificial intelligence · Digital public infrastructure · Digital public goods · Climate entrepreneurship · Climate tech

7.1 Introduction

Climate action is entering a new phase and is shifting from pledges to implementation. Energy policy, industrial policy and trade policy are increasingly becoming intertwined with climate policy. The need for bridging the gap between the current status of mobilized finance and what is required for delivering enhanced climate ambition has been recognized. Subnational actors are increasingly becoming important players in delivering climate solutions and raising climate ambition. The clean energy transition is also emerging as a key factor in economic competitiveness. Nature and biodiversity are being recognized as enablers of impactful climate solutions. Digitalization and artificial intelligence are increasingly being deployed for innovative tools for climate action.

This chapter discusses some of the emerging frontiers in climate action. Section 7.2 identifies the drivers of the new phase of climate action and identifies 'new frontiers'. Section 7.3 deep dives into Artificial Intelligence (AI) as an

K. Narula, *The Climate Catalyst*, Advances in Sustainability Science and Technology,
https://doi.org/10.1007/978-981-92-5213-8_7

enabler for climate solutions. It discusses the dual nature of AI and the growing applications of AI for climate solutions. Key applications presented include renewable energy integration, research for carbon capture materials, electric vehicles, batteries, smart buildings, urban planning solutions, climate monitoring and weather prediction. Challenges and future directions are also discussed. Section 7.4 deep dives into digital public infrastructure for accelerating climate action. Examples of digital public goods applications include open-source climate models, climate data platforms, climate risk mapping and climate decision support tools. Challenges in integration and implementation are also discussed. Section 7.5 deep dives into the growing domain of climate entrepreneurship and discusses different types of investors in climate tech. These include individual and corporate venture capital, private equity, angel investors, sovereign wealth funds and family offices. Supporting ecosystem, enablers and future outlook for climate entrepreneurship is also presented. Section 7.6 emphasizes the need for developing a systems approach for effective climate action before concluding the chapter.

7.2 A New Phase of Climate Action

7.2.1 From Incrementalism to Systems Change

The Paris Agreement has completed its ten years, and during this time there has been significant progress on climate action. The expected temperature increase without any climate policies would have been 3.5–4.0° by the end of this century, which has now been restricted to about 2.5° due to positive climate action by different countries. However, the progress is insufficient to meet the 1.5° aspirational target and the 2.0° target enshrined in the Paris Agreement. Early climate actions focus on the adoption of low-carbon technologies and marginal improvements, which have led to a slowing down in the rate of growth of GHG emissions. However, the energy system has to move from an incremental approach to a structural transformation if the goals of the Paris Agreement have to be met. This requires global coordination and a new phase of climate action, including changes to the global economic system.

While there have been several positive developments, many climate indicators are off track and require rapid deployment of climate solutions. Despite the growth of solar installations at a Compound Average Growth Rate (CAGR) of 35% and its becoming the cheapest source of electricity production in several regions of the world, it is still not on track to meet the target of tripling renewable energy capacity by 2030. Climate finance also needs to be enhanced. While the commitment to mobilize US$300 billion per year by 2035 for developing countries was agreed in 2024, the target of mobilizing international climate finance from all public and private sources to US$1.3 trillion per year is still aspirational. Further, investments in resilience and adaptation must also increase, as every dollar invested in resilience delivers over 10 times in economic benefits within a decade. Ecosystem restoration, deforestation reduction and a pullback from the tipping points are a key requirement for a structural systems transformation.

7.2.2 A World Shaped by Overlapping Crises

Over the last few years, geopolitics has defined the speed of climate action. Countries and companies are backtracking on their climate commitments, slowing down the pace of change. Conflicts continue to grow globally and a fragile security environment has led to rerouting of climate and development finance to defense spending. The global financial order is under pressure due to geopolitical fragmentation and a multipolar world is emerging with competing spheres of influence. Trust between countries is slowly eroding due to freezing of sovereign assets and increasing use of economic sanctions for political goals. De-dollarization is picking up gradually as countries diversify away from the global reserve currency, driven by concerns about the weaponization of the dollar. The Hormuz blockade impacted global energy, commodities and fertilizer prices, which has resulted in inflation, food security and energy security concerns. Global debt increased to over 300% by early 2026, and there is an increased risk of servicing sovereign debt by several countries due to higher interest rates and currency volatility. Slower economic growth, high inflation and fiscal instability are watering down climate investment.

Protectionism is increasing for home-based industries, and trade barriers are replacing global trade and coordinated action on climate mitigation. Inequality is growing between countries as well as within countries, straining the societal fabric. Technological disruptions such as artificial intelligence (AI) and the growth of data centers are increasing the demand for energy, leading to higher use of coal and gas plants for power generation in certain countries like the U.S. Higher uncertainty has led to companies adopting a wait-and-watch approach despite the increasing evidence of growing impacts and risks of climate change. These pressures together are contributing to a degraded environment and increasing poverty, which distracts the attention of governments from climate action.

7.2.3 New Frontiers for Climate Action

New frontiers of climate action include the use of AI and advanced technology for building integrated climate solutions. Digital public infrastructure (DPI) and digital public goods (DPG) are emerging as the backbone for developing enhanced decision support tools for climate action. Climate entrepreneurs are becoming an increasingly important part of the climate solution ecosystem. Funded by a wide variety of investors, they are using innovative technologies to build and implement reliable and effective climate solutions.

7.3 AI and Climate Solutions

7.3.1 The Dual Nature of AI

7.3.1.1 Environmental Footprint

The exponential growth in AI is powered by a rapid growth in data centers, which power AI workloads. In 2024, global data center electricity consumption accounted for approximately 1.5% of global electricity use, totaling 415 TWh, with the United States (45% share) followed by China (25%) and Europe (15%). The International Energy Agency (IEA) estimates that in the 'Base Case Scenario', consumption is expected to more than double to 945 TWh by 2030, while other scenarios estimate a range between 669 and 1264 TWh. The range of uncertainty widens considerably for 2035, with a forecast of approximately 1200 TWh in the base case. Other scenarios estimate between 700 TWh (if consumption plateaus) and 1700 TWh (strong AI uptake case), which would be about 4.4% of total global electricity demand (IEA, 2025).

In 2024, global data centers were responsible for approximately 180 million tonnes (Mt) of indirect CO_2 emissions resulting from electricity consumption (excluding backup generation). This was about 0.5% of global fossil fuel combustion CO_2 emissions. By 2030, CO_2 emissions from electricity generation for data centers are projected to peak at approximately 320 Mt., which would represent 3% of global power sector emissions. By 2035, CO_2 emissions are forecasted to decline to roughly 300 Mt. in the base case, and in the high-growth scenario, emissions could reach as high as 500 Mt. (IEA, 2025).

Beyond electricity use and carbon emissions, data centers have high water consumption for cooling. In 2023, global water consumption for data centers was approximately 560 billion l per year, which is forecasted to more than double to 1200 billion l per year by 2030 (IEA, 2025). A typical 100 MW hyperscale data center in the United States is estimated to consume roughly 2 million l of water per day. The intensity of water use is much higher in warm, humid climates and the Water Use Effectiveness (WUE) for direct cooling in the Asia Pacific region can be over three times the global average, which may lead to lower water availability in water-stressed regions (IEA, 2025).

The environmental footprint of AI is also high due to the generation of electronic waste driven by rapid hardware obsolescence. Graphics Processing Units (GPUs) typically become obsolete within 4 to 6 years due to the intense pace of technological advancement. The integration of AI into personal devices is expected to accelerate electronics replacement cycles, potentially leading to a surge in near-term e-waste. Use of critical minerals in electronics also puts significant pressure on supply chains, including mining and refining, which have a significant environmental footprint.

7.3.1.2 AI as Climate Solutions Accelerator

AI can also become a climate solution accelerator by optimizing existing energy systems, reducing energy consumption across end-use sectors, and shortening the timelines for scientific innovation. It is estimated that if there is 'widespread adoption' of existing AI applications, it could lead to 1.4 Gt of CO_2 emissions reductions by 2035, which is about five times more than the projected 'Base Case' emissions from global data centers (IEA, 2025).

Use cases are rapidly emerging that use AI to optimize renewable energy integration through forecasting of wind and solar electricity generation, reducing RE curtailment. AI-based power management through remote monitoring and smart grids can unlock existing grid capacity. AI can drive significant energy savings by optimizing energy consumption in industrial processes and through energy efficiency gains. AI-driven algorithms in the transportation sector can reduce emissions from heavy-duty vehicles and AI-powered building energy management systems can be used for reducing heating and cooling consumption. Satellites and sensors powered by AI can be used to identify methane leaks in oil and gas infrastructure with high precision. Continuous monitoring and use of AI-powered systems could avoid nearly 2 Mt. of methane emissions globally at a lower cost than manual programs (IEA, 2025). AI can effectively be deployed for pattern detection in climate data and predictive analytics. Use of AI can accelerate scientific discovery and innovation, as it can compress timelines by millions of times for specific tasks such as testing of new materials, battery chemistries, CO_2 capture molecules and catalyst performance simulations.

7.3.2 AI Applications for Climate Change

There are several use cases of AI for addressing climate change. Table 7.1 shows the applications of AI specific to some of the important domains.

AI applications use several technologies and techniques such as computer vision for imagery analysis; natural language processing for analyzing documents; time series forecasting for prediction; supervised learning for pattern recognition; optimization for resource allocation; clustering for grouping profiles; reinforcement learning and deep learning.

Table 7.1 Applications of AI for addressing climate change (*Data source* [48])

Domain	AI applications
Electricity systems	• Forecasting solar and wind power production • Smart grid optimization and management • Energy storage optimization • Predictive maintenance of grid infrastructure • Detecting and preventing methane leaks from natural gas infrastructure • Modeling electricity demand and emissions • Accelerating development of nuclear fusion reactors • Improving energy access in low-data contexts
Transportation	• Intelligent transportation systems • Electric vehicle charging optimization • Autonomous vehicles for higher energy efficiency • Traffic flow optimization to reduce congestion • Passenger prediction for transit planning • Freight logistics and supply chain optimization • Vehicle platooning for fuel efficiency • Alternative fuel development (e-fuels, hydrogen, biofuels)
Buildings and urban planning	• Assessment of building energy consumption • Smart building heating and cooling optimization • Energy demand forecasting • Building retrofitting • Urban planning and infrastructure design • Precision control of HVAC systems
Industry	•. Supply chain optimization for waste reduction • Manufacturing process optimization • Predictive maintenance for industrial equipment • Reducing emissions from industrial processes • Material efficiency improvements
Agriculture and forestry	• Precision agriculture for lowering water and fertilizer use • Crop yield prediction • Satellite imagery analysis for monitoring deforestation • Forest management and reforestation planning • Pest and disease detection • Soil health monitoring • Optimizing agricultural practices to reduce emissions
Carbon dioxide removal	• Identifying sites for carbon sequestration • Monitoring stored CO_2• Optimizing direct air capture systems • Managing afforestation and reforestation
Climate modeling and prediction	• Improving climate model accuracy and resolution • Extreme weather event prediction • Climate model interpretation and uncertainty quantification • Downscaling global models for regional predictions • Analyzing climate datasets
Disaster management and adaptation	• Early warning systems for extreme weather • Hurricane, flood, and wildfire prediction • Damage assessment using satellite imagery • Infrastructure vulnerability assessment • Disaster response coordination • Climate risk modeling for insurance and finance
Social and behavioral systems	• Understanding consumer energy behavior • Personalized recommendations for reducing carbon footprint • Analyzing climate-related social media data • Identifying policy impacts on emissions • Predicting financial impacts of climate change
Data and information tools	• Processing satellite imagery • Analyzing climate reports and scientific papers • Generating actionable insights from raw climate data • Creating climate datasets for research • Improving climate data availability and accessibility

7.3.3 *AI in Renewable Energy Integration and Grid Optimization*

7.3.3.1 Wind Energy Optimization

Google operates 700 MW of wind capacity in the central United States and used Google DeepMind for improving wind energy prediction. DeepMind configured a neural network trained on weather forecasts and historical turbine data to predict wind output 36 h in advance, recommending optimal hourly grid delivery commitments. This boosted wind energy value by about 20% compared to baseline scenarios, as wind energy injection into the grid could be committed on an hourly basis and its delivery could be scheduled [34].

7.3.3.2 Solar Energy Forecasting

Open Climate Fix partnered with Google DeepMind for solar forecasting. WeatherNext AI forecasting models were used to predict weather up to 15 days ahead using machine learning (ML) on several years of solar insolation data. The PVNet model integrated satellite imagery with weather predictions for solar energy forecasts. As a result, UK forecasting accuracy increased by 40% compared to previous methods. In India, results showed a 10% reduction in large errors and a 5% reduction in mean error across a 24–48-h forecasting horizon, saving millions in costs and lowering emissions [42].

7.3.3.3 Smart Grid Management

Google grid optimization project is a partnership between PJM Interconnection (the largest grid operator in North America) and Tapestry to modernize the U.S. electricity grid. It aims to deploy AI models that can intelligently manage and streamline the process of interconnecting different power sources such as renewables, conventional generation and stored energy. AI tools will help PJM speed up project approvals, cut grid planning costs and ease renewable integration. By automating verification, unifying data and models and improving analysis of variable RE, the platform will address interconnection delays and long-standing grid expansion bottlenecks [19]. Microsoft grid optimization project in partnership with the Bonneville Power Administration (BPA) in the Pacific Northwest is leveraging AI to forecast and balance electricity demand. The project applies ML algorithms to real-time grid data, and AI tools anticipate fluctuations in RE generation to adjust load distribution. This improves grid flexibility and grid resilience while integrating RE more seamlessly into the grid.

Amazon Web Services (AWS) platform in partnership with the California Independent System Operator (CAISO), is developing an AI-based energy management platform that predicts and mitigates the risks associated with RE volatility. This helps Amazon data centers to adjust their energy usage during periods of low supply and CAISO to manage grid congestion by offering real-time insights into demand patterns.

7.3.4 *AI for Carbon Capture and Materials*

At Argonne National Laboratory (U.S. Department of Energy), AI is being used for the discovery of materials for carbon capture, specifically focusing on Metal-Organic Frameworks (MOFs). MOFs are ideal candidates for carbon capture due to their porous structures, but the number of possible molecular configurations is so vast that traditional experimental testing is impractical and time-consuming. Researchers deployed a multi-stage AI workflow and generative AI was used to build over 120,000 new MOF candidates in 30 min by assembling molecular building blocks. ML and high-throughput screening filtered these candidates, while molecular dynamic simulations evaluated their stability and chemical properties. This approach drastically accelerated the discovery timeline, allowing scientists to identify the most promising, synthesizable materials rapidly (Argonne National Laboratory, 2024).

Georgia Tech and Meta have used AI to advance Direct Air Capture (DAC) technology. DAC systems must be tailored to specific environmental conditions like humidity, airflow and temperature, which requires experimental discovery of unique MOFs that can efficiently trap carbon in diverse locations. The team created the OpenDAC dataset, featuring 40 million quantum mechanics calculations for 8400 materials. Meta's Fundamental AI Research (FAIR) team trained ML models on this database to predict how MOFs interact with CO_2, orders of magnitude faster than traditional chemistry simulations, effectively narrowing down millions of possibilities in a fraction of the time. The project identified 241 high-potential MOFs for carbon capture. Open-sourcing the data and AI models provided a global toolkit for other researchers to rapidly design and deploy DAC technologies to accelerate scalable carbon removal [30].

7.3.5 *AI in Electric Vehicles and Battery Technology*

Tesla is integrating AI to improve battery performance in Electric Vehicles (EVs), which extends the life of the vehicle. EV batteries degrade due to extreme temperatures, frequent rapid charging, and inefficient energy management. Earlier, onboard vehicle management systems did not adapt to individual driving habits, which resulted in reduced range and shorter battery lifespans. Tesla has integrated an ML-powered battery management system that analyzes real-time data such as ambient temperature, battery usage patterns, and battery charging history. A trip planner feature is also included for optimal route and charging. The system dynamically adjusts charging speeds and energy distribution based on predictive analytics of the vehicle's specific needs. AI-optimized charging strategy for high-speed charging leads to improved thermal management, lower maintenance costs, and extends the life of the EV power cells. This AI-driven approach enhances battery health, and Tesla batteries have reported retaining over 90% capacity even after extended usage [28].

EVE-Ai technology developed by Electra Vehicles is used to enhance the performance of Tesla's Cybertruck. The platform uses AI to enhance battery performance, predict energy consumption, and promote safety. It employs predictive analytics in the battery management system, which continuously monitors millions of data points, including charging cycles, individual cell health, and external climate conditions. ML models 'pre-condition' cells dynamically to adjust energy consumption during off-roading to minimize thermal stress and chemical wear. It is reported that the platform can prolong battery lifespan by up to 40% by utilizing predictive maintenance and real-time driving range estimates have improved by as much as 20% [36].

Companies in the EV industry like Tesla, BYD, Rivian, Volkswagen, and General Motors are using AI to automate production lines and develop driver assistance systems. AI is being used in production to automate manufacturing. AI and ML powered robots assemble cars with precision and help speed up production while reducing defects. Computer vision powered systems enable detailed inspection to detect defects in real-time. Tesla's Full Self-Driving (FSD) and GM's *Cruise AV* use neural networks to process data from cameras, radar, and LiDAR sensors, which allow self-driving capabilities. AI also enhances adaptive cruise control, automatic emergency braking, and other enhanced safety features.

AI in supply chain and logistics allows demand forecasting and automated warehouses use AI-driven robots to manage inventory and dispatch orders efficiently. AI-powered logistics track shipments in real time, reducing delays and optimizing distribution networks. BYD is also using AI to enhance the efficiency of the *Blade Battery*, handle vehicle assembly, and in driver assistance features. Rivian is using AI-powered adaptive suspension, which adjusts in real-time for better off-road driving capability, predictive analytics, and fleet monitoring. Volkswagen and General Motors are also integrating AI to advance battery efficiency, self-driving vehicles, factory automation, optimizing battery recycling and reducing emissions in manufacturing [17].

7.3.6 AI in Smart Buildings and Cities

7.3.6.1 Building Energy Management

Building Management Systems (BMS) are deploying AI-based tools to dynamically adjust operational parameters instead of relying on fixed time-based operation schedules, leading to significant energy savings. Smart buildings use connected sensors, Internet of Things (IoT) devices, and AI in BMS to control heating, cooling, lighting, ventilation, air filtering and fire safety systems. Automated systems deploy predictive mechanisms to control the building utilities based on weather forecasts, occupancy patterns, and energy prices. The BMS also uses real-time data and learning algorithms to integrate occupant feedback, which improves comfort, increases energy efficiency, and lowers operational costs through responsiveness to changing conditions [44].

7.3.6.2 Smart City Infrastructure

Singapore is integrating AI for traffic management and uses real-time data to monitor traffic flow, predict congestion patterns, and operate traffic signals. This improves travel efficiency and reduces emissions by minimizing stop-and-go traffic, resulting in 20% lower peak-hour delays and improving average rush-hour speeds by 15%. AI is also being deployed in public transport systems to improve predictive maintenance for buses, monitoring train operations in real time, optimizing travel routes and transport schedules, resulting in a 25% increase in public transport ridership, a 15% decrease in waiting times and 10% lower emissions [21].

Barcelona is deploying AI to improve various public services. AI-assisted platforms analyze infrastructure data to predict repairs for roads and public utilities. ML algorithms analyze urban mobility patterns and pollution levels to implement environmental policies. AI is also used for planning waste collection. AI-powered sensors in trash bins analyze filling patterns and GPS-enabled waste collection vehicles are being used to optimize waste collection routes, reducing the fuel consumption of waste management fleets.

7.3.6.3 Digital Twins for Urban Planning

Barcelona is using AI for data-driven urban planning. The city developed a 'local digital twin' which is a digital replica of Barcelona powered by AI and ML. This system integrates datasets from IoT sensors, satellite imagery, and maps to simulate urban scenarios such as predicting how closing a specific street to vehicular traffic will impact air pollution and noise before implementing physical changes. The local digital twin and AI models are used to assess the '15-min city' model criteria to identify where public amenities or green spaces are lacking. This has improved public transport efficiency and optimized easy access to amenities. The Barcelona city council is also using an AI-powered employment platform for citizens, which matches employers to potential workers based on skills and delivers a tailored matchmaking service [45].

A digital twin is being used to model the impact of storms on Galveston Island, a city on the Texas Gulf Coast. The model provides insights to better understand how city planning and infrastructure development decisions impact its resilience to flooding. The digital twin can simulate the effects of flooding events under different hazard response scenarios to inform city planners about the impact of such events and enable better urban planning [40].

7.3.7 *AI for Climate Monitoring and Prediction*

7.3.7.1 Methane Detection

AI is being deployed for methane detection and addressing methane leaks, which can occur across remote infrastructures including pipelines, storage tanks, and wellheads. Manual monitoring is ineffective and error-prone, which makes precise identification of the leak and its quantification difficult. AI serves to fuse and interpret data from a multi-tiered technology stack. AI integrates data from satellites (MethaneSAT, Sentinel-5P, GHGSat), drones, and ground-based IoT sensors with meteorological data to provide a unified view of methane emissions. Models like Stanford's *GasNet* use convolutional neural networks to distinguish methane plumes from background noise in thermal imagery with up to 97% accuracy. AI algorithms pinpoint the exact source and estimate the volume of leaks and can predict potential equipment failures, allowing for planned maintenance. AI also helps in optimized sensor placement, anomaly detection and automated compliance and reporting. This has reduced the time between leak occurrence and repair, enhanced operational efficiency and lowered the cost of interventions to address methane leaks [6].

7.3.7.2 Weather and Climate Forecasting

AI is being deployed for improved predictions for extreme weather events. Google DeepMind's *WeatherNext 2* is using deep learning to provide faster and more accurate global forecasts. Existing weather forecasting models use complex physical equations to simulate the atmosphere, which requires high supercomputing power and computational time. As extreme weather events become more frequent due to climate change, timely, high-resolution, and cost-effective weather predictions have become critical. DeepMind developed *GraphCast*, an ML model based on Graph Neural Networks (GNNs) that was trained on historical global weather data. It processes weather variables across a high-resolution 3D grid of the Earth's atmosphere and identifies patterns in historical data rather than solving equations. The model is able to generate a 10-day forecast in less than a min, significantly outperforming traditional supercomputer-based weather prediction systems. The impact of *WeatherNext* has been high in terms of accuracy as it has outperformed leading climate models on over 90% of predicted variables, including wind, temperature, and pressure. Being eight times faster than earlier AI models, it allows meteorologists to run 'ensemble' forecasts to predict rare but catastrophic events with greater confidence. It also allows state-of-the-art cyclone predictions up to 15 days in advance and enables better disaster preparedness [34].

7.3.7.3 Wildfire Management

AI is being used for the early detection and management of wildfires, which have become increasingly frequent and intense due to an increase in global warming. Traditional fire detection relies on human sightings or satellite imagery that may be detected very late, making rapid intervention nearly impossible. AI-powered camera networks and sensor arrays have been deployed on mountaintops to continuously scan the horizon. AI algorithms that have been trained on large datasets of images of smoke analyze these feeds in real-time to identify wildfire smoke. AI is also used to model the propagation of fire by integrating topography, wind speeds, and vegetation moisture levels, predicting the growth of wildfires. Deployment of the technology has led to a drastic reduction in response time and allows firefighters to suppress small fires before they evolve into mega-fires. This has led to protection of lives and property, lower environmental impacts, and prevents the loss of forest-sequestered carbon and biodiversity [50].

7.3.8 Challenges and Future Directions

Deploying AI for climate solutions has some challenges that must be considered and balanced against operational risks.

7.3.8.1 Technical Challenges

Data availability and quality for predicting accurate and reliable outcomes is a concern. AI models require significant data volumes for training and sometimes have limited data. Integrating AI with existing industrial systems, including electricity grids or analog networks, is difficult unless legacy systems are upgraded. Real-time detection requires high processing power for 'edge computing', which is limited by hardware, battery life and costs in remote environments. Interoperability of systems due to non-harmonized standards and cybersecurity threats continues to present significant technical challenges.

7.3.8.2 Ethical and Social Considerations

High entry costs for AI infrastructure risk widening the gap between technologically advanced nations, developing countries and Least Developed Countries (LDCs). Open access models attempt to close this gap and offer equitable access to AI-driven technologies. However, skills gap, digital connectivity and poor infrastructure continue to persist within societies. Algorithmic bias needs to be addressed as models trained on selective data may generate solutions that are inequitable, do not consider underrepresented and vulnerable indigenous communities or may include gender biases. Data privacy issues also need to be addressed as the AI systems have access to sensitive personal data which could be leaked or hacked. Automation in certain high-employment sectors such as agriculture or industry can displace low-skilled workers, which may require social safety nets and retraining programs to ensure a just transition.

7.3.8.3 Policy and Governance Needs

Regulatory frameworks are required for governance of AI and its applications. Lack of standardized frameworks across different systems is also hindering the integration of AI with existing systems. Standardized metrics and protocols for open-source data sharing are required for enhancing international collaboration and facilitating technology transfer. Legal frameworks also need to be developed to assign responsibility for unforeseen impacts of AI deployment, such as in automated self-driving cars. Transparency and accountability must be considered in the design of AI systems and their applications.

7.3.8.4 Future Directions

Apart from technology development, using AI for building autonomous systems, developing cost-effective models which have low environmental impact and improving accuracy and speed of response, some other actions could define the future directions of growth in AI. Task forces for the use of AI in climate-critical sectors need to be established formally and informally. One such example is the Climate Change AI (CCAI), an organization composed of volunteers from academia and industry who work on tackling climate change through advancing impactful work at the intersection of climate change and ML [10]. Governments and philanthropists need to fund interdisciplinary research to bridge AI and climate domains. A part of climate funding should also be routed for the deployment of energy-efficient AI, developing use cases for innovation in AI for advancing climate action and climate impact assessments for AI systems. AI literacy programs must be developed for policymakers and industry. Cross-disciplinary education and training programs are required for building skills and capacity for best practices for responsible AI deployment. Infrastructure needs to be developed, especially in the Global South, for enabling access to AI resources and models.

7.4 Digital Public Infrastructure and Digital Public Goods for Climate Solutions

Digital Public Infrastructure (DPI) refers to the foundational society-scale digital systems and technologies that support the delivery of public and private services and are designed to be open, inclusive, interoperable and accessible. These systems serve as building blocks for the common good and facilitate economic growth, social development, and good governance. DPIs include digital identity systems, electronic payment infrastructures and data exchange platforms [38]. DPI is characterized by its public benefit and societal scale, but it can be led by the public sector, the private sector, or through public-private partnerships (PPPs).

Digital Public Goods (DPGs) are open-source digital resources such as software, datasets, artificial intelligence models, standards and content that are freely available, adhere to privacy and legal best practices and do no harm [53]. They function as shared infrastructure, enabling governments, businesses, and individuals to accelerate digital transformation for public benefit. Like other public goods, DPGs are non-rivalrous (do not diminish if others use them) and non-excludable (cannot prevent access).

Building blocks are software codes, platforms and applications that are interoperable, provide a digital service at scale, and can be reused for multiple use cases and contexts. They serve as a component of a larger system or stack of technologies and can be combined to form a DPI. Building blocks may be open source or proprietary and therefore are not always DPGs. They can be used to facilitate the delivery of digital public services via functions, which may include registration, scheduling, ID authentication, payment, data administration, and messaging [20]. DPGs and building blocks serve as the tools to build DPI and work synergistically.

Digital Public Infrastructure (DPI) is the shared backbone of digital systems and functions as an intermediate software layer that sits above physical infrastructure like communication networks, broadband and servers and below the layer of specific service applications. DPGs are the tools and applications that run on the backbone of DPI. DPGs provide the open standards that allow different components of a DPI to interact with each other seamlessly. When they provide a generic service at scale, DPGs may be considered building blocks and they may be implemented as a part of a country's digital public infrastructure.

7.4.1 *Digital Stack*

In the context of the digital infrastructure to advance climate action, a stack refers to a modular, layered digital architecture where each layer provides services for the layer above it. The lower layers provide foundational services and higher layers provide user-facing functionality. These are designed to work together through standard interfaces like defined protocols and Application Programming Interfaces (APIs).

Typically, the infrastructure layer at the bottom provides cloud computing, mobile towers, data centers, broadband connectivity, networking hardware and IoT sensors. The platform layer, which sits on top of the infrastructure layer, provides databases, digital identity management, payment systems and registries. The middleware layer consists of APIs, interoperability tools and messaging systems that connect platforms and applications. The application layer provides specific end-user solutions such as MRV tools, decision-support systems, carbon credit registries, or climate-smart agriculture apps. The top layer consists of user interface layer, which provides user-facing apps, web portals and public dashboards through which diverse stakeholders interact with the system.

7.4.2 DPI and DPGs for Climate Action

The digital stack is critical for climate action because it provides the foundational backbone and open architecture needed to scale climate solutions rapidly. While DPI establishes the underlying backbone, DPGs allow coordinated data sharing and deliver key capabilities such as open-source climate models, MRV platforms, risk mapping tools, and climate decision-support applications, which provide actionable insights for timely interventions. Open standards, plug-and-play ability, and no proprietary licensing enable governments and communities to deploy verifiable climate solutions for monitoring, emissions tracking, and early warning systems.

Modern digital stacks can play an important role in enabling large-scale climate mitigation, adaptation, and resilience. Climate solutions such as accurate GHG emissions tracking, flooding assessment, analyzing urban heat islands, real-time tracking of renewable energy production, and decentralized carbon finance can be implemented. Adaptation and resilience solutions include predictive modeling for disaster risk assessment, early-warning systems, hazard mapping, climate-smart agriculture, disaster response, resilient infrastructure planning, and real-time monitoring of vulnerable communities.

7.4.3 Examples of DPG Applications

7.4.3.1 Open-Source Climate Models and Analytics

Open-source climate models allow simulation of complex systems without proprietary software and prohibitive costs.

(a) CLIMADA (CLIMate ADAptation).

CLIMADA is a framework for climate risk assessment and evaluation of adaptation options which helps to analyze the impacts of natural hazards and explore adaptation strategies. The DPG integrates hazard data, exposure and vulnerability data to provide insights into potential socioeconomic, human, and ecological impacts of climate-related events. Beyond risk assessment, CLIMADA offers tools to evaluate the effectiveness of various adaptation measures through simulation of solutions for reducing risk and enhancing resilience [25].

(b) Oasis Loss Modeling Framework.

The *Oasis Loss Modeling Framework* provides an open-source platform for developing, deploying and executing catastrophe models. It provides access to over 90 models to simulate how different policies can mitigate financial losses from climate disasters and a toolkit for developing, testing and deploying catastrophe models [41].

(c) Climate TRACE.

Climate TRACE provides an open-access inventory for tracking the source of GHG emissions in real-time, which allows transparent, accurate, and accessible emissions data for targeted climate action. It utilizes satellite imagery, ML and AI to monitor emissions from individual facilities, cities, and countries in near real-time. It tracks emissions from all sectors, including industry, transportation, agriculture, and energy, allowing the development of data-driven decisions for mitigation strategies [15].

(d) Coupled Model Intercomparison Project.

The Coupled Model Intercomparison Project (CMIP) is a global effort to improve the understanding of climate change through the use of coupled climate models. The models simulate the interactions between multiple Earth system components such as the atmosphere, oceans, land and cryosphere. Using the enhanced understanding of Earth's climate system, it produces climate change trajectories and estimates of temperature, precipitation, sea level rise and extreme weather events based on different GHG emission scenarios. This data is used for informing policy makers to develop strategies to adapt to or mitigate the effects of climate change, including ecosystem management, disaster planning and infrastructure resilience [59].

(e) OpenClimate.

The *OpenClimate* project aims to create a 'Climate Internet' - a digitally integrated, open-source climate accounting platform to overcome the fragmentation in climate reporting. It provides a platform that links and harmonizes existing climate data from diverse sources such as IoT sensors, satellite imagery, and national databases. It uses blockchain and distributed ledger technology to prevent double counting of carbon offsets or emission reductions and tracks actions from cities, regions, and companies, which allows aggregation of data in a bottom-up manner. It also tracks the flow of finance and automates rewards for positive climate outcomes.

The climate portal collects information on the Earth system state from Intergovernmental Panel on Climate Change (IPCC) reports, climate models and global emission inventories. It integrates different registries operating at the national, subnational and non-state actor level. It integrates climate action certification information from monitoring, reporting, and verification (MRV) and certification systems. Data on climate markets, including emission records from the production of commodities and emission offsets, and information from markets operating in different jurisdictions is integrated. Climate finance data from multiple donors, multilateral development banks (MDBs) and data on project finance is also collected. The entire data is integrated through smart contracts which automate the platform [43].

7.4.3.2 Measurement, Reporting and Verification Platforms

DPG-based MRV platforms play a central role as they can enable robust tracking of emissions reduction, transparency in carbon markets and tracking of results-based finance mechanisms. As these are open tools, they lower the entry barrier for smallholders and developing countries, which do not have to rely on expensive proprietary MRV systems. Platforms like *OpenMRV* frameworks and open satellite-based monitoring systems are being applied in forestry, agriculture and energy to verify results at lower cost.

(a) OpenMRV.

OpenMRV is a knowledge platform designed to help countries and organizations measure, report, and verify forest-related data and processes [32]. It was launched by the Global Forest Observation Initiative (GFOI) and aims to bridge the technical gap in climate action by providing countries with the resources needed to operationalize their own forest MRV processes. The platform is organized into three core pillars. The measurement pillar provides technical guidance on estimating emissions and removals from forests by using activity data (area of land-use change) and emission factors (carbon per unit of area). It provides information on image classification and land-cover mapping using Google Earth Engine (GEE), System for Earth Observations, Data Access, Processing and Analysis for Land Monitoring (SEPAL), and Quantum Geographic Information System (QGIS), which are open-source tools for geospatial analysis and land monitoring. The reporting pillar provides guidance on how to transparently communicate measured data to international bodies, like the UNFCCC, using Standard Operating Procedures (SOPs). The verification pillar delivers guidance on establishing the reliability and accuracy of reported data and provides tools for Quality Assurance/Quality Control (QA/QC) and accuracy assessments. It is linked to GFOI resources like *REDDcompass* [31], which is used for system design and Collect Earth Online [27], which is a satellite image viewing and interpretation system.

(b) Climate Action Data Trust.

Climate Action Data Trust (CAD Trust) is a decentralized, open-source metadata platform designed to link and harmonize carbon credit data from registries worldwide to avoid double counting, and build confidence in carbon markets through improved transparency [8]. The open-source metadata system uses blockchain technology to create a decentralized record of carbon market activity. It provides the digital backbone for the next generation of carbon markets under Article 6 of the Paris Agreement. It creates a secure, immutable record of carbon units to prevent double counting and establishes a common data taxonomy so that different national and independent registries like Verra, ACR or Gold Standard can be harmonized and can be interoperable.

The platform operates as a decentralized metadata layer and uses blockchain technology to store a shared record of carbon market activity. A standardized framework supports complex reporting, including retirement tracking, Geographic Information System (GIS) data, and multi-methodology project support. The public data dashboard provides a user-friendly interface that aggregates data from connected registries, allowing users to search, filter, and compare carbon projects and units across different jurisdictions. It provides registry interoperability, which enables national and independent registries to connect via APIs, allowing for real-time data synchronization and cross-registry reconciliation. It also provides support for compliance reporting and auditing by providing a verifiable single source of truth for the lifecycle of carbon offsets.

(c) National Carbon Credit Registry.

National Carbon Credit Registry (NCCR) is a DPG developed by the UNDP that provides a secure, transparent technical infrastructure for helping countries to operationalize carbon markets and enabling Article 6 implementation [54]. It provides an interoperable system for managing and tracking the entire lifecycle of carbon credits and carbon offsets, from issuance to retirement. By providing the software as open-source code, it reduces the cost and technical barriers for developing countries to build national carbon registries from scratch. Namibia, Ghana, Côte d'Ivoire, Costa Rica, and Sri Lanka have been among the early adopters of the NCCR.

The platform serves as an administrative and transactional hub that manages every stage of a carbon credit, from project registration and issuance to transfer and final retirement. Key features include data collection and management, facilitating transactions and credit transfers, handling billing and accounting, and maintaining accurate records. It utilizes an immutable blockchain-based ledger to ensure that every transaction is secure and automatically assigns unique serial numbers to carbon credits, which can be tracked globally. It provides a common platform for different stakeholders, such as Designated National Authorities (DNAs) who regulate the registry, approve projects and credits, project developers and independent certifiers who verify the actual emission reductions. The platform provides interoperability and is designed to link seamlessly with other global systems, such as the Climate Action Data (CAD) Trust and national MRV tools.

7.4.3.3 Open Data for Risk Mapping and Adaptation Planning

Open climate datasets provide the database needed to identify climate hotspots, which can be used to design climate-resilient solutions. Global datasets on floods, heat stress, drought, and sea-level rise enable local authorities to assess climate vulnerabilities, which help to guide urban planning and disaster preparedness.

(a) Platform for Real-time Impact and Situation Monitoring.

Platform for Real-time Impact and Situation Monitoring (PRISM) is a climate risk monitoring system that integrates geospatial data on hazards including droughts, floods, tropical storms, and earthquakes, along with information on socioeconomic vulnerability, to inform disaster risk reduction and social assistance programmes [58]. It aims to empower governments with information on climate risk to support decision-making, such as where to direct resources to reach populations most in need of assistance. It brings together national disaster management organizations, national hydro-meteorological services and ministries such as agriculture, health, and social welfare to collectively monitor risks, prioritize responses and inform programmes and policies. PRISM visualizes data generated through satellites and sensors on the ground, which includes measurement of vegetation conditions, levels of precipitation, and temperature. It provides interactive maps and charts, climate risk monitoring indicators and impact indicators based on known vulnerabilities and exposure to hazards and the populations at risk from them. PRISM mitigates the impact of climate-driven hazards by presenting decision-makers with the most up-to-date risk and impact analytics information for climate-informed decision-making in the agriculture sector, disaster preparedness, disaster response, disaster recovery and adaptive social protection.

(b) ClimateEngine.

ClimateEngine is a high-performance web application that provides on-demand processing and visualization of massive climate and satellite datasets [12]. ClimateEngine tools use Google Earth Engine for on-demand processing of satellite and climate data in a web browser and feature on-demand mapping of environmental monitoring datasets, such as remote sensing and gridded meteorological observations. It democratizes access to big data for climate and environmental monitoring by removing technical barriers, as it allows users to analyze Earth observation data collected from 'Landsat', and 'Moderate Resolution Imaging Spectroradiometer' (MODIS) without needing to code or download files. It provides decision support insights to address risks related to drought, agriculture, wildfire, and water use by converting raw satellite data into actionable maps and charts.

ClimateEngine functions as a cloud-based middleware for Earth observations and allows comparisons of global climate impacts in real-time for monitoring planetary health. It allows on-demand mapping as users can generate customized maps and time-series charts to calculate anomalies, trends, and percentiles for variables like temperature, precipitation, and vegetation health. The platform provides a comprehensive data library with instant access to decades of satellite imagery (Landsat,

Sentinel) and climate reanalysis data (ERA5, GridMET, CHIRPS) in one place. Advanced analytics is provided through built-in tools for calculations (e.g., the evaporative demand drought index), for vegetation monitoring through tracking crop health and biomass production using the rangeland analysis platform and for performing statistical significance testing and trend analysis on climate data. APIs allow automation of workflows and the integration of ClimateEngine's data into external dashboards or corporate risk-assessment tools such as *SpatiaFi* to deliver insights about the interconnections between planetary change and economic activities [11].

(c) Copernicus Climate Data Store.

The Climate Data Store (CDS) is provided by Copernicus Climate Change Service (C3S) on behalf of the European Union [18]. It is designed as a one-stop shop for providing open access to high-quality climate datasets and data types (satellite, in-situ and models) that were previously fragmented across different sources. CDS supports adaptation and mitigation functions as it provides the evidence base needed for decision-making in the face of climate change. It functions as a broker and cloud-processor, acting as a middleman to translate users' requests into a series of commands that are sent to various global data repositories. The data is processed in the cloud and the final result is returned in the form of a map, chart, or subset of data, which makes it highly efficient and fast. The platform is built on a distributed cloud-based infrastructure with four core components. The searchable data catalogue provides a comprehensive library containing historical climate data records, real-time satellite observations and weather station data and seasonal forecasts and long-term climate model projections. The CDS toolbox provides an online software environment that allows users to write code to filter, analyze, and visualize data directly on the CDS servers. The application gallery is a collection of pre-built 'climate apps' that provide ready-to-use indicators such as river discharge, heat stress maps and solar radiation for analysis. The web-based API allows developers to automate data retrieval and integrate CDS information into their external software or workflows.

(d) Global Flood Database.

The Global Flood Database is a comprehensive satellite-based resource that provides openly accessible satellite-derived flood extent maps since the early 2000s, covering major flood events worldwide [16]. It aims to improve global flood risk management by providing reliable flood history for regions that lack ground-based sensors or historical records and quantifying the intersection between flood extents and human settlements to highlight population vulnerability. It is used for enhancing the understanding of flood risk patterns and informing adaptation strategies in vulnerable regions. It assists governments, humanitarian organizations, and insurers in planning for and mitigating the impacts of climate-driven disasters.

The platform serves as an interactive data portal and provides an interactive flood explorer tool, which is a map-based interface to view the flood extent and duration. Satellite-derived analytics provide information on the inundation extent by mapping the exact footprint of floodwaters and inundation duration. It also provides a socio-economic overlay and integrates the Global Human Settlement Layer (GHSL) to estimate the impact of the flood event on people. Unlike traditional flood models that predict flooding based on rainfall and terrain, this model uses direct observation from space. By processing over 15 years of daily satellite imagery, it has estimated that the proportion of the population exposed to floods is increasing ten times faster than previously estimated.

Using the flooding database, Floodbase quantifies uncertainty between flood models for risk pricing to determine parametric insurance based on satellite-verified flood levels rather than manual damage assessments [16].

(e) Climate Hazard Maps.

Climate Hazard Maps can be created for individual countries. A specialized geospatial data portal has been developed by the United Nations Environment Programme (UNEP) to support Iraq's climate resilience and adaptation planning. The portal's main objective is to provide evidence-based, geographically visualized data on climate risks to support the Iraq National Adaptation Plan (NAP). It operates as a climate GIS data portal and translates complex global climate models into localized, easy-to-understand maps. By focusing on 'first-order' (physical) and 'second-order' (socio-economic) risks, it helps to identify priority areas for infrastructure investment, water management, and disaster risk reduction [55].

It aims to support decision-making for government officials and development partners to integrate climate risks into national and regional policy, as it shows where and how natural hazards are expected to change in magnitude and frequency across Iraq. It serves as an interactive mapping and data repository hub that provides national-scale maps to visualize climate hazards. Six specific climate hazard indices are assessed including heat index, dry spell, warm and dry days, standardized and precipitation intensity index, and sea level rise. The portal hosts maps based on Coordinated Regional Climate Downscaling Experiment (CORDEX) data, which provides high-resolution regional climate projections specifically in the Iraqi context. The model allows users to view potential future climate scenarios (e.g., 2030, 2050, 2080) based on different GHG emission pathways. The interactive web map app allows users to overlay different datasets to perform their own spatial analysis. The open data repository also provides access to downloadable datasets, documentation, and technical reports such as the climate risk assessment for Iraq.

7.4.3.4 Decision-Support and Policy Simulation Tools

DPGs also provide support for decision support and policy design. Open decision-support and policy simulation tools help governments to test alternative policy options before implementation. These tools embed best-practice methods in simulation software which help to inform the design of actionable policy decisions.

(a) En-ROADS Climate Solutions Simulator.

En-ROADS (Energy Rapid Overview and Decision Support) is an interactive global climate simulator which allows users to explore how different policy, technological, and economic shifts can affect global warming and the climate system [13]. The simulator allows policymakers to test combinations of climate policies (e.g., carbon pricing, renewable energy, energy efficiency, etc.) to observe their impacts on global temperature, emissions, and energy systems through the year 2100.

It provides a transparent sandbox for testing the impact of policies before being implemented and provides a fast, scientifically rigorous tool that runs in real-time, allowing for 'what-if' scenario testing which can be used during live workshops or negotiations. En-ROADS works as a global aggregated model which does not simulate specific countries but looks at the world as a whole. It provides a simple browser interface which allows policymakers to use it as a tool for simulating climate action. It uses about 30 policy sliders to set energy supply, electrification and energy efficiency penetration, deployment of technological solutions like DAC and several other parameters. The 21,000-equation model works behind the scenes to generate results such as expected warming by 2100 in real time. It also provides outputs such as energy costs, GDP impact, carbon tax revenue, air quality (PM2.5), sea-level rise and crop yield changes. Detailed user controls are available to adjust the specific tax amount or the year of policy implementation, providing a fast and scientifically rigorous tool to test policy scenarios.

(b) Climate Policy Radar.

Climate Policy Radar is an AI-powered, open-source platform designed to map, track, and analyze the global climate policy landscape, which provides a searchable database of 30,000+ national climate laws, policies and official submissions like Nationally Determined Contributions (NDCs) and Global Stocktake documents [14]. It aims to democratize climate policy data to support evidence-based decision-making and organizes fragmented and siloed information on climate policies and international agreements into a single, unified hub. This also allows comparison of legal frameworks and climate interventions across different jurisdictions. The platform uses advanced Natural Language Processing (NLP) tools and AI to turn documents into structured data. By providing open data and open-source code, it creates a DPG which allows practitioners to build their own tools on top of its high-quality data pipeline.

(c) Climate Impact Explorer.

The Climate Impact Explorer (CIE) is an interactive data visualization tool which aims to provide a comprehensive, scientifically rigorous bridge between global temperature increase and their specific, localized physical impacts [9]. The tool aims to assess physical risks under different economic and emission scenarios and provides an analysis dashboard to present impact indicators such as projected changes in crop yields, frequency and intensity of heatwaves, droughts and extreme precipitation, impacts on labor productivity due to heat stress and freshwater availability indices. Different policy scenarios can be compared and the impact of different levels of warming (1.5 °C, 2 °C, 2.5 °C, 3 °C) can be assessed. The CIE operates on a 'nested model' architecture, which takes global climate sensitivity data and downscales it to produce localized impacts. Unlike models that focus only on the atmosphere, the CIE model focuses on the biophysical response by isolating how climate change alone affects specific metrics while other socio-economic conditions remain unchanged. Thus, it shows the direct consequences of increasing global warming regardless of future population moves or technological changes.

(d) Energy Access Explorer.

The Energy Access Explorer (EAE) is a DPG that provides an interactive geospatial platform to assist energy planners in identifying optimal sites for renewable energy deployment based on local demand and resources [60]. The platform supports a bottom-up approach and aims to enable data-driven energy planning by linking energy supply with socio-economic demand and providing granular, ground-level data to locate high-demand areas for expanding energy access. The tool functions as a dynamic GIS that allows users to perform complex spatial analysis directly in a web browser without specialized software. Users can overlay and weight more than 50 different datasets, such as population density, poverty rates, solar/wind potential, and proximity to the power grid, to perform a Multi-Criteria Analysis (MCA) for generating customized maps to identify priority areas. It maps the supply of energy, which includes renewable resource availability, current infrastructure such as transmission lines and substations and the demand, which tracks socio-economic indicators such as health facilities, schools, agricultural productivity and household wealth. The backend automatically harmonizes and integrates new satellite imagery and public datasets, reducing the technical burden on local planning agencies.

(e) Catastrophe Simulation.

Catastrophe Simulation (CatSim) is a specialized modeling tool designed to help policymakers, particularly in disaster-prone and developing nations, assess and manage the financial risks associated with natural catastrophes and develop public financing strategies for disaster risk management [35]. The model aims to identify financing gaps and to determine if a government has sufficient immediate capital to fund relief and reconstruction after a major disaster such as floods, earthquakes or hurricanes. It helps users to evaluate the most cost-effective mix of pre-disaster options like insurance and post-disaster financing options like international aid or loans. It also provides a scientific basis for mainstreaming disaster risk management into long-term national development and climate adaptation plans. CatSim functions as a stochastic financial stress-test model and predicts the economic aftershock of a climate disaster. By treating disaster risk as a contingent liability, the model shifts the focus from reactive disaster response to proactive financial resilience.

The risk assessment module assesses the value of assets exposed based on users defined parameters (frequency, intensity) for natural hazards. It estimates the potential economic losses based on the probability of different disaster magnitudes. The financial resilience analysis module analyzes the government's ability to reallocate budgets, use reserve funds, or increase taxation following an event and calculates the fiscal gap between available resources and projected reconstruction costs. The policy simulation module can be used to simulate the impact of various financial tools, such as catastrophe bonds, contingent credit lines, reserve funds, and traditional insurance and to evaluate how these financial choices affect long-term indicators like national debt, GDP growth, and social stability. An interactive interface allows policymakers to visualize risk through maps and charts in real-time. The model has been used to help launch the Caribbean Catastrophe Risk Insurance Facility (CCRIF) and has assisted Mexico in issuing its first-ever catastrophe bond.

7.4.4 DPI and DPGs for Enhanced Climate Ambition

DPI is emerging as the backbone of data transparency, which can strengthen climate governance and help in scaling up climate solutions. DPGs also democratize climate solutions by improving accessibility and reducing costs through the use of open-source technologies, architecture and data. Digital payment infrastructure built on DPI can accelerate delivery of targeted subsidies, insurance payouts and incentives with minimal transaction costs and leakages, leading to greater financial inclusion in the green transition. Digital climate platforms can provide integrated solutions for mobility, energy planning, financial and administrative services, optimizing resource use. Digital systems also lead to faster implementation and can improve the speed of climate policy execution.

7.4.5 Challenges in Integration and Implementation

The digital divide continues to be a structural barrier in implementing innovative climate solutions. Lack of internet access, reliable infrastructure and poor digital literacy limit the reach of digital climate solutions and tools to the most vulnerable communities, which may reinforce existing inequalities. Interoperable and standardized digital protocols are essential for using shared infrastructure for MRV, improving accuracy and traceability in data reporting, but face persistent challenges. Data privacy concerns are increasing due to cybersecurity vulnerabilities and poor data governance safeguards. Maintenance of digital systems requires continuous attention and technical support, which has higher associated costs. Data quality, inconsistencies in data reporting and reliability also need to be overcome for implementing climate solutions.

7.5 Climate Entrepreneurship

Climate entrepreneurs address climate challenges through developing new business models, creating innovative products and services. They focus on long-term solutions which have a strong component of social and environmental responsibility. Growing sectors for climate entrepreneurship include renewable energy and storage, sustainable agriculture, food systems, waste management, circular economy, transportation systems including electric mobility and shared mobility, alternative proteins, carbon removal, energy efficiency and climate fintech. Key characteristics of climate entrepreneurs include risk-taking, innovation, collaboration and technology integration.

7.5.1 The Investment Landscape

Global climate tech funding peaked in 2022 with investments close to US$76 billion [39]. For 2025, the total annual funding is projected at US$53 billion, an increase compared to US$49 billion in 2024. The number of total deals fell from 6495 in 2022 to a projected 3751 in 2025, showing that capital is flowing into fewer companies. Grant funding plunged 51% while private equity deals fell 18% year-on-year (YoY) in 2025. Early-stage climate tech faced increasing pressures and pre-seed and seed activity hit a five-year low in 2025, with 50% fewer companies progressing from Seed to Series A funding compared with 2022. Companies progressing from Series A to Series B also declined by 80% versus 2022 and the Series A-to-B transition took six months longer on average.

Over 70% of climate tech funding flows into climate mitigation. Batteries, solar, and grid technologies dominate and account for 55% of the funding share. However, adaptation and resilience are becoming increasingly important investment themes. Adaptation's share of climate tech equity funding rose from 5% to 8.2% in one year and has doubled since 2021. Resilience technologies, which have strong investment momentum, include wildfire management and drought management, where funding has increased by 20–25% YoY.

Funding in AI increased from $12B in 2023 to $14B in 2025 and AI-enabled companies captured about 28% of total Climate Tech equity funding in 2025. AI is now seen as an important infrastructure and data centers attracted close to $2 billion in funding. Other strong AI-enabled growth areas include battery optimization, solar forecasting, precision agriculture and climate-risk analytics. Funding for industrial sectors increased nearly 30% YoY with growth in green steel, smart manufacturing, nuclear energy, and advanced materials. Technologies with stronger industrial demand and policy backing are progressing faster. Grid infrastructure and industrial heating have shown relatively strong commercialization success, while green hydrogen, electro fuels, DAC, and long-duration storage continue to lag.

Overall, investments in climate tech are slowing down and maturing into projects that are deployment ready and have execution capability. Capital is now flowing into projects which demonstrate commercial readiness, scalability, and deployment certainty, with investors choosing companies with proven demand and reliable supply chains.

7.5.2 *Leading Climate Tech Unicorns*

Climate tech unicorns are private companies that develop innovative climate solutions and have a valuation of over US$1 billion. They are leaders in climate innovation and provide solutions across different domains. Table 7.2 shows the top ten clean tech unicorn startups. The list is not definitive as valuations change over time and there may be differences in how companies are classified due to multiple revenue streams.

7.5.3 *Investor Types*

Investors are a central player in the climate tech startup ecosystem. They play a different role in financing climate startups, as they have higher risk tolerances and time frames. Some of the important investor types are shown below.

7.5.3.1 Venture Capital

Venture capitalists (VCs) provide funding in exchange for equity and target companies that have high-growth potential in the early to growth stages. They are one of the largest funders in the climate tech space and drive the growth of technological innovation, with typical investment ranges from US$1 M to US $50 M. VCs have high risk appetite, expect high returns and exit their investment within 5–10 years through mergers and acquisitions (M&A) or initial public offerings (IPOs).

Breakthrough Energy funds companies with the potential to cut at least half a gigaton of GHG per year across the five sectors of manufacturing, electricity, agriculture, transportation and buildings [4]. *Energy Impact Partners* is another VC firm that invests across venture, growth, private equity and credit stages to help companies scale and deploy clean energy technologies and sustainability innovations [23].

7.5.3.2 Private Equity

Private equity (PE) investors focus on late-stage companies that have a proven product or service. These companies are established and require capital to scale operations for expanding market share, or to enter new regions. PE investment is less risky, as the companies are mature and the focus is on operational efficiencies and rapid growth. The ticket size for investments is typically from US$10 M to US$500 M. Risk appetite is moderate, as companies have stable revenue models. PE firms target returns when companies have public offerings or secondary sales to other firms.

The 'Carlyle Group' is a global PE firm that invests in the energy transition through integrated infrastructure and energy platforms to scale mature, capital-intensive solutions. Its infrastructure platform provides funding to renewable and sustainable

Table 7.2 Top 10 clean tech unicorn companies (*Data source* [26])

Company	Approx. valuation (as on Dec 2025)	Headquarters	Founding year	Funding	Primary focus	Last funding
Northvolt *	$12B	Stockholm, Sweden	2016	$13.8B	Sustainable lithium-ion batteries for EVs and energy storage	Private equity
Octopus energy group	$9B	London, UK	2015	$2.9B	RE supply, kraken technology platform	Private equity
Pivot bio	$2B	Berkeley, CA, USA	2011	$616.9 M	Agricultural microbes for natural nitrogen fertilization	Series D
Rad power bikes	$2B	Seattle, WA, USA	2007	$329.3 M	Electric bike manufacturer	Series D
Arcadia	$2B	Greenwood Village, CO, USA	2014	$575.5 M	Climate software and data platform for grid decarbonization	Series E
Prometheus	$2B	San Mateo, CA, USA	2018	$150 K	Zero-net-carbon fuels from atmospheric CO_2 capture	Series B
Climeworks	$1B	Zürich, Switzerland	2009	$946.1 M	Direct air capture technology for CO_2 removal	Private equity
Svante	$1B	Burnaby, BC, Canada	2007	$505.3 M	Carbon capture for heavy industries (cement, steel)	Venture round
Gradiant	$1B	Woburn, MA, USA	2013	$442.4 M	Advanced water and wastewater treatment solutions	Series D
Palmetto	$1B	Charleston, SC, USA	2009	$2.2B	Clean energy platform for home electrification	Private equity

* Northvolt declared bankruptcy in 2025

energy companies for projects on EV infrastructure, battery storage, solar development, and sustainable infrastructure [5]. 'Blackstone' is another PE firm investing in clean tech and energy transition, with a goal to invest close to US$100 billion in climate change solutions over a decade. *Blackstone Energy Transition Partners* is Blackstone's energy-focused PE business and has invested over US$25 billion of equity globally within the energy industry [3].

7.5.3.3 Angel Investors

Angel investors are high-net-worth individuals who provide early-stage funding in the seed or pre-seed stage to tech startups. These investors have a strong commitment to addressing climate change and offer capital and mentorship to companies. Very early-stage companies or startups with innovative climate solutions are targets for angel investors, as these companies are not established and ineligible for funding from other investors. The investment size typically ranges from US$50 K to US$1 M. The risk appetite is very high as companies may have unproven technologies and unclear business models. Angel investors may also support through follow-up funding rounds that lead to later-stage investments.

Well-known angel investors include John Doerr, Peter Kellner and Scott Sandell. John Doerr, through the venture capital firm *Kleiner Perkins*, invests in climate tech and the portfolio includes climate-focused companies like *Watershed* and *Rondo Energy*. Peter Kellner, through the company *Richmond Global LLC*, focuses on the seed stage, with a typical investment range of US$200 K to US$2 M. Scott Sandell invests in later-stage rounds, specifically Series A and B, through the company *New Enterprise Associates* in the cleantech sector with notable investments in companies like *VAST Data* and *Transfix* [46].

7.5.3.4 Corporate Venture Capital

Large corporations have their own venture capital arms to invest in climate tech startups. Corporate Venture Capital (CVC) funds invest in climate tech companies that complement or enhance their existing business lines. Through these investments, CVC firms aim to drive innovation in technologies and align their operations with corporate sustainability commitments. CVC investment size depends on the stage of funding and ranges from US$1 M to US$50 M. The risk appetite is moderate to low, as they seek strategic value rather than high growth. CVC firms often integrate the funded company into their own operations in a few years or may sell the company.

Shell Ventures, established in 1996, is the CVC arm of petroleum conglomerate 'Shell', which invests and partners with companies to commercialize innovative businesses. It invests in early-stage and growth-stage clean tech companies dealing in power, mobility, emission management and digital technologies. They contribute with a typical initial investment of US$2-5 M, with a total of US$15-22 M over the lifecycle of an investment [49]. *ENGIE New Ventures*, founded in 2014, is the corporate venture arm of the global energy provider 'ENGIE'. It invests in innovative solutions to accelerate the energy transition. With investments of over €250 M, the fund invests in cleantech technology startups that complement existing activities and resources to spur internal innovation within ENGIE. It has an initial investment size of €1-5 M and has invested in over 25 startups [24].

7.5.3.5 Impact Investors

Impact investors seek to generate financial returns and positive environmental and social impacts. They invest in companies working at the intersection of sustainability and business, such as water conservation, sustainable agriculture, and low-carbon technologies. The investment size ranges from US$500 K to US$50 M depending on the project. Their risk appetite is moderate and investors are ready to accept lower financial returns if the social impact of the project is high. Impact investors are focused on long-term returns and exit the investment when the firm has a public offering or strategic acquisition.

Generation Investment Management, a firm co-founded by former US vice president Al Gore, focuses on investments in companies addressing global environmental challenges. The company seeks to deliver long-term, risk-adjusted investment returns, positive impact, and advocates for the adoption of sustainable investing by the wider market [29]. The *Rockefeller Foundation* is a philanthropy that focuses on impact investing to advance global climate solutions. It has committed over US$1 billion from 2023–2028 to transform energy, food, health, and finance systems. It deploys patient and risk-tolerant catalytic capital to bridge funding gaps in climate finance. It has funded projects on clean energy access, battery storage and grid infrastructure, electric transportation, sustainable land use and agriculture, carbon capture, nature-based solutions, climate-smart infrastructure and renewable energy deployment in emerging markets. It acts as a liaison between investors and entrepreneurs while experimenting with innovative financing solutions to mobilize large-scale private sector capital toward clean energy infrastructure and climate resilience [47]. It has provided funding for the *Global Energy Alliance for People and & Planet* (US$500 million initial commitment) and the *Battery Energy Storage Systems Consortium*, which aims to bring electricity to over 3 billion people.

7.5.3.6 Sovereign Wealth Funds

Sovereign wealth funds (SWFs) are government-owned funds that deploy capital for climate tech and sustainable infrastructure projects. They typically fund large-scale renewable energy projects and sustainable infrastructure projects with investment over US$100 M. The risk tolerance is low and they invest in high-impact climate solutions. Funds are deployed for the long term, and exits are timed strategically over decades.

Examples include Norway's Government Pension Fund Global (GPFG), the world's largest sovereign wealth fund with over US $1.9 trillion in assets (as of June 2025), which actively invests in renewable energy and clean tech. Abu Dhabi Investment Authority (ADIA), established in 1976, is one of the world's largest sovereign wealth funds, which invests globally in sustainable infrastructure and renewable energy projects and energy storage. It has a strategic alliance with Masdar, Abu Dhabi's leading renewable energy company and supports the Mubadala Investment Company in financing large-scale climate and cleantech initiatives. Through its research unit, ADIA Labs Initiative, it focuses on data science, AI, ML, and quantum computing to inform global trends on energy transition [1].

7.5.3.7 Family Offices

Family offices are privately held wealth management firms that manage the assets of ultra-high-net-worth individuals or families. Several family offices invest in the climate tech sector due to personal commitments of the individuals. They invest in both early-stage and established companies across the climate tech ecosystem, with typical investment ranging from US$1 M to US$50 M. They have a moderate to high-risk appetite and prefer long-term investments with strategically timed exits through M&A or IPOs.

Grok Ventures, established in 2016, is Mike Cannon-Brookes' family office. It pledged a US$1.5 billion investment to address climate change over 2021–2031. It plans to invest across the entire climate capital stack from seed-stage startups to large renewable energy infrastructure. It invested US$200 million in the Infradebt Energy Transition Fund (2022) for 1.5-2GW battery projects and has investments in *Sun Cable,* which is the Australia-Asia solar transmission project. *Capricorn Investment Group* is a part of the Jeff Skoll family office, founder of eBay. It has US$5 billion in assets in direct management and around US$3.5 billion in partnerships. Through the *Technology Impact Fund I* and *Growth Fund,* it has invested in climate tech moonshot projects such as *Helion Energy,* which develops small-scale fusion reactors and *Form Energy,* which develops long-duration iron-air battery technology.

7.5.4 Support Ecosystems and Enabling Initiatives

Several initiatives specialize in creating a supportive ecosystem for climate entrepreneurs and advancing climate innovation. Global climate entrepreneurship platforms and coalitions convene startups, investors, policymakers, corporates, and funders to build a supportive ecosystem. Climate accelerators, incubators and venture builders provide hands-on technical, commercial, and fundraising support to companies. Public-sector and multilateral initiatives provide support by shaping market conditions, creating early demand for services and contributing to funding to unlock scale. Initiatives include World Bank/IFC Climate Innovation Programs, UNFCCC

Climate Technology Centre and Network (CTCN) and the EU Innovation Fund. Some of the other important initiatives are mentioned ahead.

(a) Coalition for Climate Entrepreneurship.

The 'Coalition for Climate Entrepreneurship' (CCE) is a public-private partnership designed to catalyze climate innovation in developing economies by nurturing emerging and market-viable climate solutions [57]. CCE works with several partners to support climate-focused startups and provide entrepreneurial support through nine global hubs in Morocco, Ghana, Kenya, Colombia, Costa Rica, Guatemala, Jamaica, Vietnam, and in the U.S. Silicon Valley. Major forms of support include accelerator programs which provide structured business development pathways from ideation to implementation, technical skill-building for climate entrepreneurs, mentorship, access to capital, knowledge sharing and business development tools, among others. CCE has leveraged over US$2.5 M from more than 30 private sector partners and has supported several market-ready, climate-focused ventures.

(b) UNICEF Climate Ventures.

UNICEF 'Climate Ventures' is a five-year initiative (2025–2030) which aims to bridge the gap between early-stage climate innovation and large-scale deployment of climate solutions. It intends to support up to 60 ventures which focus on child-critical climate challenges, including strengthening community resilience. The initiative acts as a catalytic connector and strategic incubator that provides structured access to UNICEF country offices, government partnerships and market entry support, including providing capital. This includes a supportive ecosystem tailored for growth-stage startups with mentorship from experts to co-create national integration plans and navigate new markets and opportunities for connecting entrepreneurs with a global pipeline of donors [56].

(c) Climate Action Innovation Zone.

The 'Climate Action Innovation Zone' is independently organized alongside the UN Climate Change Conference and focuses on mobilizing the global innovation ecosystem. It facilitates collaboration between stakeholders and aims to transform high-level climate ambition into practical implementation. Through events like the 'Sustainable Innovation Forum' and the 'Climate Implementation Summit', it facilitates matchmaking between investors and climate-tech startups. By focusing on accelerating the deployment of climate technology solutions, the annual gathering helps to reduce the time for creating a global market for climate innovation [7].

(d) Elemental Impact.

'Elemental Impact' is a climate accelerator which focuses on bridging the transition from early-stage innovation to commercial-scale deployment for climate technologies. It addresses market failures associated with First-of-a-Kind (FOAK) and early commercial projects. It provides several services for climate entrepreneurs. This includes catalytic capital and innovative financial tools to de-risk early projects and unlock follow-on investment. It also supports projects with technical expertise to ensure that projects are deployment-ready in areas of policy engagement, workforce development and capital coaching. The accelerator has over 160 portfolio companies and 150+ projects deployed globally and operates a US$94 M venture fund, *Earthshot Ventures.* It has catalyzed over US$11.5 billion in additional funding through its network of over 40 philanthropic and government funders [22].

(e) Third Derivative.

'Third Derivative' (D3) is a climate tech accelerator founded in 2020 by the Rocky Mountain Institute (RMI) and New Energy Nexus. It aims to catalyze markets by integrating startups into a unified ecosystem of capital and industry expertise. Third Derivative acts as a systemic orchestrator and a hub to bridge the disconnect between startups, risk-averse corporates and fragmented investment landscapes in the climate tech sector. It accelerates the commercialization of breakthroughs in hard-to-abate sectors like heavy industry, grids and storage, and clean power. D3 offers services such as *Venture Capital Integration,* where startups receive US$100 k + in initial funding and are plugged into a network of financers for follow-on investment. It also supports startups with corporate matchmaking to secure pilot projects and first-customer contracts, apart from giving technical support to refine business models. Through the innovation cohorts, D3 runs specialized programs, such as the *Clean Mobile Power Initiative* and *First Gigaton Captured*, which focus on solving specific, high-stakes emission challenges. D3 support has led to reduced time-to-market for innovative decarbonization technologies, lower risk for investors and corporates and has mobilized capital for high-impact climate solutions [52].

(f) Greentown Labs.

'Greentown Labs' is a climate tech incubator which supports the climate innovation ecosystem by bridging the gap between laboratory research and commercial scale for early-stage climate startups. It provides hardware-centric infrastructure and resources to lower the barriers to hardware innovation by providing prototyping services, which include shared labs and machine shops, allowing startups to build and test physical products. It has structured accelerator programs such as *Go Build* and *Go Energize* that connect startups with corporate partners and facilitate access to climate-focused venture capitalists and angel investors. It also offers legal, HR, and business software tools and guidance to startups. Greentown Labs has supported over 500 companies that have collectively raised more than US$5 billion in funding for decarbonizing key sectors including agriculture, buildings, electricity and manufacturing.

(g) SOSV Climate Tech 100.

The 'SOSV Climate Tech 100' is an ecosystem initiative by SOSV, a global venture capital firm investing in deep tech. It serves as an early-stage investor and institutional co-founder, providing the missing link of lab space and engineering expertise. It runs specialized programs such as *HAX* (hard tech) to identify and de-risk projects. It also provides the high-risk, pre-seed capital and technical infrastructure necessary to commercialize innovations that are too complex for traditional VCs. It provides seed capital and a six-month residency with access to labs, machine shops, and prototyping equipment. Startups are given access to on-site experts for engineering and design support as well as supply chain experts to help scale manufacturing. SOSV hosts VC-founder matchups to provide access to a network of investors for follow-on funding. Their portfolio includes companies focused on green steel, carbon mitigation and methane-reducing enzymes and has successfully catalyzed several climate solutions [51].

7.5.5 Future Outlook

Climate entrepreneurship and climate tech investing are maturing and entering a selective phase, which is evidenced by the slowing global investments. The market opportunity remains untapped, but capital and talent are shifting to competitive sectors. Global spending on clean energy technologies crossed US$2 trillion in 2024 and the potential to integrate innovation in technology, business models and service delivery are growing.

Apart from the supply of renewable energy, advanced energy distribution technologies, higher efficiency in end-use sectors and fuel switching technologies in the energy system, there is a clear business case for using data analytics. Development in commercial models will enable scaling up of climate solutions. Energy systems integration offers a good opportunity for optimization of energy supply, demand and utilization. Flexibility in electricity grids is becoming valuable and startups that integrate energy storage, demand response, supply forecasting and variable loads will emerge as winners. Companies that help utilities and system operators optimize assets and manage grid congestion will have more opportunities.

Climate adaptation and resilience is also emerging as an investable space. Extreme heat, floods, droughts, and wildfire risk are creating demand for climate risk analytics and insurance products. Companies specializing in integrating early-warning systems and developing resilient infrastructure will find new markets opening up, especially in the Global South where countries are exposed to higher climate risk.

Emerging trends in climate tech include advancement and application of AI and digital technologies for compliance reporting, advancing carbon markets, hydrogen, EVs, precision technologies and scaling decarbonization, especially in hard-to-abate sectors. Market opportunities in climate-enabling infrastructure such as MRV platforms, carbon accounting for supply chains and providing climate risk data for insurers are growing. Innovation to provide energy access, cooling, E-mobility, and agricultural services will continue to grow but would require context-specific solutions that leverage technology with local service delivery leading to result-oriented outcomes.

7.6 Building Positive Momentum for Climate Action

Effective climate action requires a system approach as energy, environment, economy and social systems are deeply intertwined. Climate solutions and policies often have competing tradeoffs which must be balanced with the objectives of economic growth and affordability. Climate mitigation, adaptation, and resilience need to be addressed simultaneously and resources must be allocated judiciously. Integrated approaches can unlock cross-sector synergies and stronger coordination between ministries, regulators, local governments, and private sector actors is essential for the effectiveness of climate action. Systems thinking also improves policy durability and effectiveness of climate action in the face of evolving socio-political dynamics and climate shocks.

Universal approaches have limited effectiveness. Country-specific climate solutions need to be developed and integrated climate pathways must account for national circumstances and capabilities. Fit-for-purpose climate solutions which account for local context are essential for effective policies. Low-carbon resilient infrastructure must be prioritized across all sectors and centralized planning needs to be complemented with local implementation.

Positive climate momentum builds on early successes, which are visible in local economic benefits from successful implementation of projects. Aligning vision and climate ambition with execution is also essential for successful climate action. Clear roadmaps, strengthening institutional coordination and improved implementation capacity build credibility and confidence for businesses to invest in climate solutions. Climate action is slowly being mainstreamed but needs to be embedded in all economic models, risk assessment frameworks and investment decisions. The process of implementing policy shifts is equally important and must be accompanied by consultation and a participatory process. Policies must be fair and inclusive as they strengthen political stability and increase acceptability. Cross-sector collaboration and partnerships which adopt a whole of society approach accelerate momentum and can help scale solutions more rapidly. The narrative is equally important and climate action must be framed around opportunities for economic growth, improved health outcomes, increased jobs, and affordability.

To be effective, climate action must move beyond siloed thinking. Climate action must recognize that incremental changes and piecemeal approaches are insufficient as the gap between climate commitments and climate action continues to widen. Integration of technology, finance, governance, and social innovation is essential for effective climate action. Further delays in addressing climate issues will increase reliance on climate trajectories that overshoot the set thresholds and increase climate risks. It is therefore essential that climate action must be scaled rapidly globally so that it is beneficial for the entire society.

7.7 Conclusion

Climate action is entering a new phase and is in the process of being mainstreamed in decision-making. Driven by systems thinking, digitalization and development of digital public infrastructure, climate entrepreneurship is reshaping how climate solutions are deployed and scaled. AI is emerging as an enabler for climate solutions and has several innovative applications for addressing climate change, such as renewable energy forecasting, materials discovery, electric vehicles, batteries, smart buildings, urban planning and climate monitoring. Open climate data, interoperable digital platforms, digital public goods, and climate decision-support tools can improve transparency, strengthen climate decision-making and enable the development of climate solutions. Nevertheless, there are some challenges, such as the growing energy demand of AI systems, data privacy issues and concerns about responsible governance, which need to be addressed to leverage the full potential of AI for scaling climate solutions.

Climate entrepreneurship is emerging as a driver of commercialization of climate solutions. A functional ecosystem comprising entrepreneurs, venture capital, private equity, sovereign wealth funds, family offices, accelerators, incubators, and corporate players is required to drive climate tech-driven solutions. Climate technologies, digital infrastructure, financial systems and climate governance frameworks must be designed as interconnected components to achieve long-term effectiveness. AI can significantly accelerate both climate mitigation and climate adaptation solutions. However, future climate solutions require systems thinking and coherent solutions at scale rather than isolated interventions.

Successful and effective climate action lies at the intersection of climate governance, policy, finance, markets and technology. The success of forthcoming climate solutions will be measured by their ability to accelerate progress towards global net-zero emissions and climate-resilient development.

References

1. Abu Dhabi Investment Authority (n.d.) Homepage. https://www.adia.ae/. Accessed 13 12 2025
2. Argonne National Laboratory (2024) Argonne Scientists Use AI to Identify New Materials for Carbon Capture.https://www.anl.gov/article/argonne-scientists-use-ai-to-identify-new-materials-for-carbon-capture. Accessed 18 11 2025
3. Blackstone (n.d.) Blackstone Energy Transition Partners. https://www.blackstone.com/our-businesses/blackstone-energy-transition-partners/. Accessed 13 12 2025
4. Breakthrough Energy (n.d.) Programs. https://www.breakthroughenergy.org/programs. Accessed 08 12 2025
5. Carlyle (n.d.) Global Private Equity. https://www.carlyle.com/our-firm/global-private-equity. Accessed 08 12 2025
6. Chubb (n.d.) Using AI to Pinpoint and Predict Methane Leaks. Chubb Methane Resource Hub. https://about.chubb.com/content/dam/chubb-sites/chubb/about-chubb/citizenship/environment/pdf/methane-ai.pdfAccessed 20 11 2025

7. Climate Action (n.d.) Innovation Zone. https://events.climateaction.org/innovation-zone/. Accessed 13 12 2025
8. Climate Action Data Trust (n.d.) Homepage. https://climateactiondata.org/. Accessed 01 12 2025
9. Climate Analytics (n.d.) Climate Impact Explorer.https://climate-impact-explorer.climateanalytics.org/. Accessed 05 12 2025
10. Climate Change AI (n.d.) About. https://www.climatechange.ai/about. Accessed 25 11 2025
11. Climate Engine (2023) Introducing SpatiaFi. https://climateengine.com/introducing-spatiafi/. Accessed 01 12 2025
12. Climate Engine (n.d.) About. https://www.climateengine.org/about. Accessed 01 12 2025
13. Climate Interactive and MIT Sloan (n.d.) En-ROADS Climate Solutions Simulator. https://en-roads.climateinteractive.org/scenario.html?v=25.11.0. Accessed 05 12 2025
14. Climate Policy Radar (n.d.) Homepage. https://www.climatepolicyradar.org/. Accessed 05 12 2025
15. Climate TRACE (n.d.) Explore. https://climatetrace.org/explore. Accessed 28 11 2025
16. Cloud to Street/Floodbase (n.d.) Global Flood Database. https://global-flood-database.cloudtostreet.ai/. Accessed 05 12 2025
17. CodeDesign.ai (2026) AI-Driven Innovations in Electric Car Manufacturing. https://codedesign.ai/blog/ai-driven-innovations-in-electric-car-manufacturing/. Accessed 18 03 2026
18. Copernicus Climate Change Service (n.d.) Climate Data Store. https://cds.climate.copernicus.eu/. Accessed 05 12 2025
19. Data Center Frontier (n.d.) Google's AI-Powered Grid Revolution: How Data Centers Are Reshaping the U.S. Power Landscape. https://www.datacenterfrontier.com/energy/article/55282719/googles-ai-powered-grid-revolution-how-data-centers-are-reshaping-the-us-power-landscape (Accessed: 18 November 2025)
20. Digital Public Goods Alliance (n.d.) DPI, DPG and Building Blocks: Definitions. https://www.digitalpublicgoods.net/DPI-DPG-BB-Definitions.pdf (Accessed: 25 November 2025)
21. EARTHDAY.ORG (2025) Smart cities, green futures: how artificial intelligence is powering urban sustainability. https://www.earthday.org/smart-cities-green-futures-how-ai-is-powering-urban-sustainability/Accessed 20 11 2025
22. Elemental Impact (n.d.) Mission and Impact. https://elementalimpact.com/mission-and-impact/ Accessed 28 06 2026
23. Energy Impact Partners (n.d.) Partners. https://www.energyimpactpartners.com/partners/Accessed 08 12 2025
24. ENGIE New Ventures (n.d.) About ENGIE New Ventures. https://www.engieventures.com/about/ (Accessed: 13 December 2025)
25. ETH Zurich, Weather and Climate Risks Group (n.d.) CLIMADA—Climate Risk & Adaptation Platform. https://climada.ethz.ch/ Accessed 25 11 2025
26. Failory (2025) The Full List of 15 CleanTech Unicorn Startups (2026). Updated 9 December. https://www.failory.com/startups/cleantech-unicorns(Accessed: 20 June 2026)
27. FAO Open Foris (n.d.) Collect Earth. https://www.collect.earth/ Accessed 30 11 2025
28. Gail (2023) Unlocking the power: Tesla's AI revolution in EV battery technology. Xcelerate Auto https://www.xcelerateauto.com/blog/AI-and-battery-autonomy-Teslas. Accessed 18 11 2025
29. Generation Investment Management (n.d.) Our Firm. https://www.generationim.com/our-firm/ (Accessed: 15 December 2025)
30. Georgia Tech News Center (2024) Georgia Tech and meta create massive open dataset to advance AI solutions for carbon capture. https://www.gatech.edu/news/2024/05/02/georgia-tech-and-meta-create-massive-open-dataset-advance-ai-solutions-carbon(Accessed: 18 November 2025)
31. Global Forest Observations Initiative (n.d.) REDDcompass: Actions. https://www.reddcompass.org/actions/en-3/filter(Accessed: 30 November 2025)
32. Global Forest Observations Initiative and World Bank (n.d.) OpenMRV. https://openmrv.org/home (Accessed: 30 November 2025)

33. Google DeepMind (n.d.) Machine Learning Can Boost the Value of Wind Energy.https://deepmind.google/blog/machine-learning-can-boost-the-value-of-wind-energy/ Accessed 17 11 2025
34. Google DeepMind (n.d.) WeatherNext 2. https://deepmind.google/science/weathernext/ (Accessed: 20 November 2025)
35. IIASA (n.d.) Catastrophe Simulation (CATSIM). Available at: https://iiasa.ac.at/models-tools-data/catsim. (Accessed: 08 December 2025)
36. Interesting Engineering (2025) New AI brain boosts tesla Cybertruck range by 20%, Battery Life by 40% CES 2025. https://interestingengineering.com/ces-2025/tesla-cybertrucks-battery-lifespan-improvementAccessed 18 11 2025
37. International Energy Agency (2025) Energy and AI. 'AI and Climate Change' section. Paris: IEA. https://www.iea.org/reports/energy-and-ai/ai-and-climate-change. Accessed 15 11 2025
38. International Telecommunication Union (n.d.) Digital Public Infrastructure. ITU-D Technology.https://www.itu.int/en/ITU-D/Technology/Pages/dpi.aspx. Accessed 25 11 2025
39. Net Zero Insights (2025) State of Climate Tech 2025. https://stateofclimatetech.com/wp-content/uploads/2025/12/NZI_2025-State-of-Climate-Tech-Report.pdf. Accessed 20 06 2026
40. Nira R (2021) 'Digital Twins' of Texas coastal communities could shed light on resiliency. Texas a&M University. College of Architecture. https://stories.tamu.edu/news/2021/10/08/digital-twins-of-texas-coastal-communities-could-shed-light-on-resiliency/. Accessed: 20 11 2025
41. Oasis Loss Modelling Framework (n.d.) Homepage.https://oasislmf.org/. Accessed 28 11 2025
42. Open Climate Fix (2025) AI-Based Weather Forecasting Is Enabling the Renewable Energy Transition.https://www.openclimatefix.org/insights/ai-based-weather-forecasting-is-enabling-the-renewable-energy-transition. Accessed 17 11 2025
43. OpenEarth Foundation (n.d.) Homepage. https://www.openearth.org/. Accessed 30 11 2025
44. Penn State Institute of Energy and the Environment (2025) How Smart Buildings Use AI to Cut Energy Use and Improve Health. https://iee.psu.edu/news/blog/how-smart-buildings-use-ai-cut-energy-use-and-improve-health. Accessed 20 11 2025
45. Popelka, S. (2023) AI for sustainability in Barcelona. Urban AI, Medium. https://medium.com/urban-ai/ai-for-sustainability-in-barcelona-cc5135d252d0. Accessed 20 11 2025
46. Rho (n.d.) Funding Your Clean Energy Startup: 9 VCs to Know. https://www.rho.co/blog/angel-investors-for-cleantech. Accessed 13 12 2025
47. Rockefeller Foundation (n.d.) Homepage. https://www.rockefellerfoundation.org/. Accessed 13 12 2025
48. Rolnick, D., Donti, P.L., Kaack, L.H., Kochanski, K., Lacoste, A., Sankaran, K., Ross, A.S., Milojevic-Dupont, N., Jaques, N., Waldman-Brown, A., Luccioni, A.S., Maharaj, T., Sherwin, E.D., Mukkavilli, S.K., Kording, K.P., Gomes, C.P., Ng, A.Y., Hassabis, D., Platt, J.C., Creutzig, F., Chayes, J. and Bengio, Y. (2022) 'Tackling Climate Change with Machine Learning', ACM Computing Surveys, 55(2), 42. https://dl.acm.org/doi/10.1145/3485128. Accessed 15 11 2025
49. Shell (n.d.) About Shell Ventures. https://www.shell.com/what-we-do/technology-and-innovation/innovate-with-shell/shell-ventures/about.html. Accessed 13 12 2025
50. Simon, J. (2024) How companies are using artificial intelligence to tackle global warming. NPR. https://www.npr.org/2024/01/02/1218677963/ai-climate-change-solutions-fires-lithium-methane. Accessed: 20 11 2025
51. SOSV (n.d.) Programs. https://sosv.com/programs/. Accessed 13 12 2025
52. Third Derivative (n.d.) Homepage. https://www.third-derivative.org/. Accessed 13 12 2025
53. UN Office for Digital and Emerging Technologies (n.d.) Digital Public Infrastructure. https://www.un.org/digital-emerging-technologies/sites/www.un.org.digital-emerging-technologies/files/general/Digital_Public_Goods_Summary_PDF.pdf. Accessed: 25 11 2025
54. UNDP (n.d.) National Carbon Credit Registry. https://nationalcarboncreditregistry.com/index.html. Accessed: 01 12 2025
55. UNEP Iraq (n.d.) Climate GIS Data Portal. https://home-unep-iraq.hub.arcgis.com/pages/climate-hazard-maps. Accessed: 05 12 2025
56. UNICEF (n.d.) The Next Wave of Climate Entrepreneurship. Available at: https://www.unicef.org/innovation/stories/next-wave-climate-entrepreneurship.Accessed: 13 12 2025

57. US Department of State (n.d.) Coalition for Climate Entrepreneurship *(CCE)*. 2021–2025 Archive. https://2021-2025.state.gov/coalition-for-climate-entrepreneurship-cce/ Accessed: 13 12 2025
58. WFP Innovation (n.d.) PRISM. https://innovation.wfp.org/project/prism (Accessed: 01 12 2025)
59. World Climate Research Programme (n.d.) CMIP—Coupled Model Intercomparison Project. https://wcrp-cmip.org/cmip-overview/. Accessed 28 11 2025
60. World Resources Institute (n.d.) Energy Access Explorer. https://www.energyaccessexplorer.org/ (Accessed: 08 12 2025)

GPSR Compliance

The European Union's (EU) General Product Safety Regulation (GPSR) is a set of rules that requires consumer products to be safe and our obligations to ensure this.

If you have any concerns about our products, you can contact us on ProductSafety@springernature.com

In case Publisher is established outside the EU, the EU authorized representative is:

Springer Nature Customer Service Center GmbH
Europaplatz 3
69115 Heidelberg, Germany

Batch number: 10799748

Printed by Printforce, the Netherlands